FIXED POINT THEOREMS AND THEIR APPLICATIONS

FIXED POINT THEOREMS AND THEIR APPLICATIONS

Ioannis Farmakis

Martin Moskowitz

City University of New York, USA

 **World Scientific**

NEW JERSEY · LONDON · SINGAPORE · BEIJING · SHANGHAI · HONG KONG · TAIPEI · CHENNAI

Published by

World Scientific Publishing Co. Pte. Ltd.

5 Toh Tuck Link, Singapore 596224

USA office: 27 Warren Street, Suite 401-402, Hackensack, NJ 07601

UK office: 57 Shelton Street, Covent Garden, London WC2H 9HE

British Library Cataloguing-in-Publication Data
A catalogue record for this book is available from the British Library.

Cover: Image courtesy of Jeff Schmaltz, MODIS Land Rapid Response Team at NASA GSFC.

ISBN 978-981-4458-91-7

Printed in Singapore

Contents

Preface and Acknowledgments

Our intention here is to show the importance, usefulness and pervasiveness of fixed point theorems in mathematics generally and to try to do so by as elementary and self contained means as possible. The book consists of eight chapters.

Chapter 1, *Early fixed point theorems*, deals mostly with matters connected to the Brouwer fixed point theorem and vector fields on spheres, the Contraction Mapping Principle and affine mappings of finite dimensional spaces. As we shall see, the Brouwer theorem can be generalized in basically two ways, one leading to the Lefschetz fixed point theorem in topology (whose proof is independent of Brouwer) and the other to the Schaüder-Tychonoff fixed point theorem of analysis (which depends on the Brouwer theorem). We shall draw a number of consequences of Brouwer's theorem: among them Perron's theorem concerning eigenvalues of a "positive" operator, which itself has interesting applications to the modern world of computing such as in Google addresses and ranking in professional tennis.

Chapter 2, *Fixed point theorems in Analysis*, deals with the theorems of Schaüder-Tychonoff, as well as the those of F. Hahn, Kakutani and Kakutani-Markov involving groups of affine mappings, the latter two leading naturally to a discussion of Amenable groups.

Chapter 3 concerns *Fixed point theorems in Topology*, particularly the Lefschetz fixed point theorem, the H. Hopf index theorem and some of their consequences. As a further application we prove the conjugacy theorem for maximal tori in a compact connected Lie group. In addition, we explain the Atiyah-Bott fixed point theorem and its relationship to the classical Lefschetz theorem.

Chapter 4, *Fixed point theorems in Geometry*, is devoted first to the fixed point theorem of E. Cartan on compact groups of isometries of Hadamard manifolds and then to fixed point theorems for compact manifolds, when the curvature is negative, due to Preissmann, and positive, due to Weinstein.

Chapter 5 concerns *Fixed points of maps preserving a volume form*, which are important in dynamical systems. We begin with the Poincaré recurrence theorem and deal with some of its philosophical implications. Then we turn to symplectic geometry and fixed point theorems of symplectomorphisms. Here we discuss Arnold's conjecture, prove Poincaré's last geometric theorem and derive a classical result concerning billiards. We then turn to hyperbolic automorphisms of a compact manifold, particularly of a torus, and then to Anosov diffeomorphisms and the analogous Lie algebra automorphisms and explain their significance in dynamics. We conclude this chapter with the Lefschetz zeta function and its many applications.

Chapter 6 deals with the *Fixed point theorem of A. Borel* for a solvable algebraic group acting on a complex projective variety, the most important such varieties being the Grassmann and flag varieties. We then present some consequences of these ideas. A final section concerns a fixed point theorem giving rise to a conjugacy theorem for unipotent subgroups of real reductive linear Lie groups.

We conclude with two brief chapters. Chapter 7 deals with some connections of fixed points to number theory, group theory and complex analysis, while in Chapter 8, *A fixed point theorem in Set Theory*, we prove Tarski's fixed point theorem and, as a consequence, the Schröder-Cantor-Bernstein theorem.

As the reader can see, fixed point theorems are to be found throughout mathematics. These chapters can, for the most part, be read independently. Thus the reader has many options to follow his or her particular interests.

We thank Oleg Farmakis for his help in creating the diagrams and Hossein Abbaspour for reading an earlier draft of the manuscript and making a number of useful suggestions for improvements. Of course, any mistakes are the responsibility of the authors. Finally, we thank Konstantina and Anita for their patience.

New York, January 2013
Ioannis Farmakis, Martin Moskowitz

Introduction

As exactly a century has passed since the Brouwer fixed point theorem [22] was proved and because in some sense this result is the progenerator of some of the others, it seems appropriate to consider fixed point theorems afresh and in general. As we shall see, these are both quite diverse and pervasive in mathematics. Fixed point theorems are to be found in algebra, analysis, geometry, topology, dynamics, number theory, group theory and even set theory. Before proceeding it would be well to make precise what we mean by a fixed point theorem.

Definition 0.1.1. Let X be a set and $f : X \to X$ be a map from X to itself. A point $x \in X$ is called a *fixed point* of f if $f(x) = x$.

A fixed point theorem is a statement that specifies conditions on X and f guaranteeing that f has a fixed point in X. More generally, we shall sometimes want to consider a family $\mathcal{F}$ of self maps of X. In this context $\mathcal{F}$ is usually a group (or sometimes even a semigroup, see [82]). In this case a fixed point theorem is a statement that specifies conditions on X and $\mathcal{F}$ guaranteeing that there is a *simultaneous* fixed point, $x \in X$, for each $f \in \mathcal{F}$.

When $\mathcal{F}$ is a group G this will usually arise from a *group action* $\phi : G \times X \to X$ of G on X. We will write $\phi(g, x) := gx$. We shall assume that the 1 of G acts as the identity map of X and for all $g, h \in G$ and $x \in X$ that $(gh)x = g(hx)$. We shall call such an X, a *G-space*. Particularly, when X is a topological space and G is a topological group, we shall assume ϕ is *jointly continuous*.

1

Now, as above, when we have an action of a group G on Y, by taking for "X" the power set, $\mathcal{P}(Y)$, this induces an action of G on $\mathcal{P}(Y)$ and the fixed points of this new action are then precisely the G-invariant subsets of Y. In this way, in many contexts, fixed point theorems give rise to G-invariant sets.

We conclude this introduction with the remark (and example) that the existence of Haar measure μ on a compact topological group G can be regarded as a fixed point theorem. This is because in the associated action of G by left translation on the space $M^+(G)$ of all finite positive measures on G, left invariance simply means μ is G-fixed. Moreover, since μ is finite and can be normalized so that $\mu(G) = 1$, we can even regard it as a G-fixed point under the action of G on the *convex* (and compact in the weak* topology) set of positive normalized measures on G (see Corollary 2.3.5).

Chapter 1

Early Fixed Point Theorems

1.1 The Picard-Banach Theorem

One of the earliest and best known fixed point theorems is that of Picard-Banach 1.1.1. Either explicitly or implicitly this theorem is the usual way one proves *local* existence and uniqueness theorems for systems of ordinary differential equations (see for example [79], sections 7.3 and 7.5). This theorem also can be used to prove the inverse function theorem (see [79], pp. 179-181).

Theorem 1.1.1. *Let (X, d) be a complete metric space and $f : X \to X$ be a contraction mapping, that is, one in which there is a $1 > b > 0$ so that for all $x, y \in X$, $d(f(x), f(y)) \leq bd(x, y)$. Then f has a unique fixed point.*

Proof. Choose a point $x_1 \in X$ (in an arbitrary manner) and construct the sequence $x_n \in X$ by $x_{n+1} = f^n(x_1)$, $n \geq 2$. Then x_n is a Cauchy sequence. For $n \geq m$,

$$d(x_n, x_m) = d(f^n(x_1), f^m(x_1)) \leq b^m d(f^{n-m}(x_1), x_1).$$

But

$$d(f^{n-m}(x_1), x_1) \leq d(f^{n-m}(x_1), f^{n-m-1}(x_1)) + \ldots + d(f(x_1), x_1).$$

The latter term is $\leq (b^{n-m-1} + \ldots + b + 1)d(f(x_1), x_1)$ which is itself $\leq \sum_{n=0}^{\infty} b^n d(f(x_1), x_1)$. Since $0 < b < 1$ this geometric series converges to $\frac{1}{1-b} d(f(x_1), x_1)$. Since $b^m \to 0$ we see for n and m sufficiently large, given $\epsilon > 0$,

$$d(x_n, x_m) \leq b^m \frac{1}{1-b} d(f(x_1), x_1) < \epsilon.$$

Hence x_n is Cauchy. Because X is complete $x_n \to x$ for some $x \in X$. As f is a contraction map it is (uniformly) continuous. Hence $f(x_n) \to f(x)$. But as a subsequence $f(x_n) \to x$ so the uniqueness of limits tells us $x = f(x)$.

Now suppose there was another fixed point $y \in X$. Then $d(f(x), f(y)) = d(x, y) \leq bd(x, y)$ so that if $d(x, y) \neq 0$ we conclude $b \geq 1$, a contradiction. Therefore $d(x, y) = 0$ and $x = y$. □

We remark that the reader may wish to consult Bessaga ([11]), or Jachymski, ([60]) where the following converse to the Picard-Banach theorem has been proved.

Theorem 1.1.2. Let $f : X \to X$ be a self map of a set X and $0 < b < 1$. If f^n has at most 1 fixed point for every integer n, then there exists a metric d on X for which $d(f(x), f(y)) \leq bd(x, y)$, for all x and $y \in X$. If, in addition, some f^n has a fixed point, then d can be chosen to be complete.

As a corollary to the Picard-Banach theorem we have the following precursor to the Brouwer theorem.

Corollary 1.1.3. Let X be the closed unit ball in $\mathbb{R}^n$ and $f : X \to X$ be a nonexpanding map; that is one that satisfies $d(f(x), f(y)) \leq d(x, y)$ for all $x, y \in X$. Then f has a fixed point.

Proof. For positive integers, n, define $f_n(x) = (1 - \frac{1}{n})f(x)$. Then each f_n is a contraction mapping of X. Since X is compact, it is complete. By the contraction mapping principle each f_n has a fixed point, $x_n \in X$ and since X is compact x_n has a subsequence converging to say x. Taking limits as $n \to \infty$ shows x is fixed by f. □

We now state the Brouwer fixed point theorem.

Theorem 1.1.4. *Any continuous map f of the closed unit ball B^n in $\mathbb{R}^n$ to itself has a fixed point.*

In other words, *if one stirs a mug of coffee then at any given time there is at least one particle of coffee that is in exactly the position it started in.*

Of course, since the Brouwer theorem is stated in purely topological terms, it is actually true for any topological space X homeomorphic to B^n; so in particular, for any compact convex set in $\mathbb{R}^n$. (As an exercise the reader might want to prove this.) Notice that as compared to the Picard-Banach theorem, here we have no uniqueness. Also, the space is rather specific, for example it is compact (and convex), but the map is rather general. In Picard-Banach it is the opposite. The space is merely complete, but the map is required to have very strong properties and so we get uniqueness. In the Brouwer theorem there is no uniqueness. For example the identity map has all of B^n as fixed points.

To illustrate the Brouwer theorem in dimension one, consider a square (with boundary) S in the plane and the diagonal going from lower left to upper right. Choose two points, one on the left edge of S and one on the right. Then the result says that any continuous curve joining these points and lying wholly in the square must intersect the diagonal.

One way to prove the Brouwer fixed point theorem is by studying vector fields on Spheres (Theorem 1.2.1). In Chapter 2 we shall see that this result is also a consequence of the Lefschetz fixed point theorem and that it also implies Theorem 1.2.1.

1.2   Vector Fields on Spheres

Let X be a compact submanifold of Euclidean space $\mathbb{R}^{n+1}$. A continuous (resp. smooth) *tangential vector field* v on X is a continuous (resp. smooth) $\mathbb{R}^{n+1}$-valued function on X with the property that for each $x \in X$, $v(x)$ is tangent to X at x, that is it lies in the tangent hyperplane, $T_x(X)$. We shall call v *nonsingular* if $v(x)$ is never 0 at any point of X. For example, if $X = S^2$, the 2-sphere, considered as the surface of the earth, then the tangential component of the wind gives a tangential vector field (which one assumes is continuous, but *which may turn out to be singular*). Given a nonsingular smooth, or continuous tangential vector field v on X, we can produce a unitary continuous tangential vector field w on X, that is, one whose vectors are all of unit length, simply by normalizing v at each point: $w(x) = \frac{v(x)}{\|v(x)\|}$, $x \in X$. One checks easily that w is smooth (resp. continuous). An example of a nonsingular (unit) tangent vector field on S^1 is given by choosing any point, x_0, and placing a unit vector v_0 at x_0 tangent to the circle there. Then since any other point $x \in S^1$ can be obtained from x_0 by a unique counter clockwise rotation we just rotate v_0 by the same amount and get a unit tangent vector to x. Here we are particularly interested in the unit sphere, $X = S^n \subseteq \mathbb{R}^{n+1}$. In this case each tangent vector $v(x)$ is perpendicular to x, $x \in S^n$. That is $\langle x, v(x) \rangle = 0$ on S^n. One verifies easily that this condition is both necessary and sufficient for tangency.

The first question we address is: are there nonsingular smooth (or continuous) vector fields on S^n? When $n = 2k - 1$ is odd there always are. For example, we will show that for $x = (x_1, \ldots, x_{2k}) \in S^{2k-1}$,

$$v(x_1, \ldots, x_{2k}) = (x_2, -x_1, x_4, -x_3, \ldots, x_{2k}, -x_{2k-1})$$

is such a vector field. For v is evidently a smooth function of x. To see that it is unitary, notice $v(x_1, \ldots, x_{2k}) = x_e - x_o$, where x_e is the vector in $\mathbb{R}^{2k}$ with even subscripts x_{2i} in the odd coordinates and zeros in the even ones, while x_o is the vector in $\mathbb{R}^{2k}$ with zeros in the odd coordinates and x_{2i-1} in the even ones. Hence

$$\| v(x) \|^2 = \langle x_e - x_o, x_e - x_o \rangle = \| x_e \|^2 + \| x_o \|^2 + 2\langle x_e, x_o \rangle.$$

However, $\langle x_e, x_o \rangle = 0$ and $\| x_e \|^2 + \| x_o \|^2 = \| x \|^2 = 1$. Therefore, $\| v(x) \|^2 \equiv 1$. Moreover, one sees immediately using the fact that we are in an even dimensional Euclidean space that $\langle v(x), x \rangle = 0$. Thus v is a smooth unitary tangential vector field on X.

Notice that one cannot even write such a thing when n is even. Indeed as we shall see, when n is even there are no continuous nonsingular vector fields on X. So for example, *at any moment there must be*

a point on the earth's surface where the wind does not blow tangentially.
Viewed from the aspect of vector fields it is this fact which lies behind
the Brouwer theorem.

In this section we prove the following result, Theorem 1.2.1 (sometimes called the hairy ball theorem). Then, we shall extend Theorem 1.2.1 from the smooth to the continuous case from which we will
get the Brouwer fixed point theorem 1.1.4. Finally, from Theorem 1.1.4
as a corollary we will get the Invariance of Domain theorem 1.3.1 and
other applications.

Theorem 1.2.1. *S^n possesses a nonsingular smooth vector field if and
only if n is odd.*

Since we have shown just above that odd dimensional spheres always
have a nonsingular smooth vector field, it remains to prove that if n is
even there can be no such vector field.

Let $n = 2k$, i.e. n is even and suppose there were a smooth unitary
tangential vector field v on S^n. For $t \in \mathbb{R}$ let $F_t(x) = x + tv(x)$, for all
$x \in S^n$. We calculate $\| x + tv(x) \|$. Just as above

$$\langle x + tv(x), x + tv(x) \rangle = \langle x, x \rangle + 2t\langle x, v(x) \rangle + t^2\langle v(x), v(x) \rangle.$$

Since both $\| x \| = \| v(x) \| = 1$ and $\langle x, v(x) \rangle = 0$, we see that

$$\| x + tv(x) \| = \sqrt{1 + t^2}.$$

To complete the proof of our theorem we shall need the following two
lemmas.

Lemma 1.2.2. *If $|t|$ is sufficiently small, F_t is a $1 : 1$ mapping taking
X onto a region whose volume can be expressed as a polynomial function
of t.*

Let vol_* denote the n-dimensional volume (as opposed to the $n + 1$
dimensional volume in $\mathbb{R}^{n+1}$).

Lemma 1.2.3. *If $|t|$ is sufficiently small, the map F_t takes S^n onto the
sphere of radius $\sqrt{1 + t^2}$. Hence $\mathrm{vol}_*(F_t(S^n)) = (1 + t^2)^{\frac{n+1}{2}} \mathrm{vol}_*(S^n)$.*

These two lemmas suffice to prove Theorem 1.2.1 since when $n+1$ is odd $(1+t^2)^{\frac{n+1}{2}} \operatorname{vol}_*(S^n)$ is not a polynomial function of t, contradicting lemma 1.2.2. The reader should verify that this is so.

Proof of Lemma 1.2.2. Since X is compact and v is continuously differentiable on X by the mean value theorem there is some positive constant c so that for all $u, x \in X$, $\| v(u) - v(x) \| \le c \| u - x \|$. We know this is true for u, x in a common neighborhood. Since X is compact we can cover it by a finite number of such neighborhoods and then choose c to be the minimum of the constants associated with each of these neighborhoods.

Now let $|t| < \frac{1}{c}$. If $F_t(x) = F_t(u)$ then $\| u - x \| \le |t|c \| u - x \|$. Hence $u = x$ and F_t is $1:1$. Taking $|t|$ small so this happens and observing that the derivative $d(F_t)(x) = I + t \, dv(x)$, where I is the identity, we see the Jacobian, $\det(d(F_t)(x)) = \det(I + t \, dv(x))$, is a polynomial function of t (of degree n). Now we integrate this polynomial over X by using the change of variables formula (as we may since F_t is $1:1$) and see that $\operatorname{vol}_*(F_t(X))$ is itself a polynomial function of t (whose coefficients are the integrals over X of the coefficients of $\det(I + t \, dv(x))$. This completes the proof of Lemma 1.2.2.

Proof of Lemma 1.2.3. Consider the region R between two concentric spheres defined by the inequalities $a \le \| u \| \le b$, where $a < 1$ and $b > 1$. Extend the vector field v to R by $v(rx) = rv(x)$, where $a \le r \le b$. This also extends the map F_t to all of R because $F_t(rx) = rF_t(x)$ and moreover, using this last equation, it follows that F_t maps R onto the region between the spheres of radii $a\sqrt{1+t^2}$ and $b\sqrt{1+t^2}$.

Now by the same argument as in Lemma 1.2.2 for $|t|$ sufficiently small this extended F_t is $1:1$ on R. Therefore by the chain rule its derivative, $d(F_t)(x)$, is invertible at every point. The inverse function theorem tells us that F_t maps open sets in R to open sets in $F_t(R)$. Hence $F_t(S^n)$ is an open set in the sphere of radius $\sqrt{1+t^2}$. But this nonempty open set is also compact and therefore closed. Since spheres are connected it must be the entire sphere of radius $\sqrt{1+t^2}$, proving Lemma 1.2.3.

We remark just as with the Brouwer theorem, Theorem 1.2.1 is true for much more than just the unit sphere. In fact, it is true for anything diffeomorphic with the unit sphere. For example, it is true of an ellipsoid of any size, shape and orientation. Another example to keep in mind is a partially inflated basketball. A nonsingular smooth vector field on such a surface will become a nonsingular smooth vector field on the sphere by simply blowing up the ball. More formally, if $\Phi : X \rightarrow S^n$ is the diffeomorphism, then for each $x \in X$, $d(\Phi)_x : T_x(X) \rightarrow T_{\Phi(x)}(S^n)$ is an invertible linear map between their tangent spaces. Hence it transports a nonsingular smooth vector field on X to one on S^n.

We now extend Theorem 1.2.1 to continuous vector fields.

Corollary 1.2.4. *S^n possesses a nonsingular continuous vector field if and only if n is odd.*

Proof. Suppose an even dimensional sphere X had a continuous unitary vector field, v. By the Weierstrass approximation theorem (see e.g. [79] p. 374) there is a polynomial map $p : X \rightarrow \mathbb{R}^{n+1}$ so that $\| p(x) - v(x) \| < \frac{1}{2}$. Let $w(x) = p(x) - \langle p(x), x \rangle x$. Then this is a smooth vector field on X. Is it tangential?

$$\langle w(x), x \rangle = \langle p(x) - \langle p(x), x \rangle x, x \rangle = \langle p(x), x \rangle - \langle p(x), x \rangle \langle x, x \rangle = 0$$

since $\| x \| = 1$. Thus w is tangential. Since both v and w are tangential,

$$\langle p(x) - w(x), x \rangle = \langle p(x), x \rangle = \langle p(x) - v(x), x \rangle.$$

By the Schwarz inequality, the latter is $\leq \frac{1}{2}$. Hence, by the definition of w,

$$|\langle p(x), x \rangle| \leq \frac{1}{2}$$

everywhere on X. Therefore, $\| p(x) - w(x) \| < \frac{1}{2}$.

Also, since $\| p(x) - v(x) \| < \frac{1}{2}$, we see $\| v(x) - w(x) \| < 1$ and because v is unitary, w is never zero. Thus, w is a smooth nonsingular vector field on an even dimensional sphere, which is impossible by Theorem 1.2.1. □

1.3   Proof of the Brouwer Theorem and Corollaries

We can now prove the *Brouwer fixed point theorem*.

We will prove this by means of Corollary 1.2.4. Just as above, this theorem and its consequences apply to any space homeomorphic with B^n.

Proof. Let $B^{n+1} \subseteq \mathbb{R}^{n+1}$ denote the closed unit disk and first consider the case when S^n has even dimension. If we can prove Theorem 1.1.4 here, then the case when n is odd, say $n = 2k - 1$ can be easily handled. For any continuous map $f : B^{2k} \to B^{2k}$ without fixed points gives rise to a map $F : B^{2k+1} \to B^{2k+1}$ as follows: $F(x, x_{2k+1}) = (f(x), 0)$, where $x \in B^{2k}$. Evidently F is continuous and maps B^{2k+1} to itself. Any fixed point (x, x_{2k+1}) of F would clearly make x a fixed point of f.

We now consider n to be even and assume f is a continuous map from B^{n+1} to itself without fixed points. Since $f(x) \neq x$ the vector field $v(x) = x - f(x)$ is never zero on B^{n+1}. This vector field points *outward* in the sense that $\langle u, v(u) \rangle > 0$ everywhere on S^n, since

$$\langle u, v(x) \rangle = \langle u, u \rangle - \langle u, f(u) \rangle = 1 - \langle u, f(u) \rangle \geq 1 - |\langle u, f(u) \rangle|.$$

But by the Schwarz inequality, since $f(u) \in S^n$, $|\langle u, f(u) \rangle| \leq 1$ and is strictly less than 1 (and therefore $\langle u, v(u) \rangle > 0$) unless $f(u) = \lambda u$. Because both these vectors are in S^n we would then have $f(u) = \pm u$. However, if $f(u) = u$ this violates our assumption that f has no fixed points and if $f(u) = -u$, $1 - \langle u, f(u) \rangle = 2$.

Next we modify v to get a nonzero vector field w on B^{n+1} pointing *directly outward* on the boundary, that is $w(u) = u$ everywhere on S^n. For $x \in B^{n+1}$ set

$$w(x) = x - \frac{f(x)(1 - \langle x, x \rangle)}{1 - \langle x, f(x) \rangle}.$$

Notice that $1 - \langle x, f(x) \rangle$ is never zero. If it were, then as above, $f(x) = \lambda x$, where $\lambda \in \mathbb{R}$. Hence $1 - \langle x, f(x) \rangle = 1 - \lambda \langle x, x \rangle = 0$ so $0 < \lambda \leq 1$. If $\lambda = 1$ then as before x would be a fixed point of f. If $\lambda < 1$, then $1 - \lambda \langle x, x \rangle$ cannot be 0. Since all functions above are continuous it follows that w depends continuously on x. Evidently $w(x) = x$ whenever $\langle x, x \rangle = 1$.

We check that w is never zero. This is certainly true if x and $f(x)$ are linearly independent. If they are dependent, i.e. if $f(x) = \lambda x$, then $\langle x, x \rangle f(x) = \langle x, f(x) \rangle x$ and so if $w(x) = 0$, we conclude $x = \lambda x$ so again $\lambda = 1$ and x would be a fixed point of f.

Next we transplant w from B^{n+1} to the southern hemisphere of the unit sphere, S^n, of $\mathbb{R}^{n+1}$ by means of stereographic projection (or rather its inverse) σ from the north pole, $N = (0, 0, \ldots, 1)$. That is, $\sigma : \mathbb{R}^n \to S^n - N$, where the former is identified as the points of $\mathbb{R}^{n+1}$ whose last coordinate $x_{n+1} = 0$. The formula for σ is

$$\sigma(x) = \frac{(2x_1, \ldots, 2x_n, \langle x, x \rangle - 1)}{\langle x, x \rangle + 1},$$

where $x = (x_1, \ldots, x_n, x_{n+1}) \in S^n - N$. As is easily checked, $\| \sigma(x) \| = 1$. Since if $\frac{(2x_1, \ldots, 2x_n, \langle x, x \rangle - 1)}{\langle x, x \rangle + 1}$ were to be $(0, \ldots, 0, 1)$, $\langle x, x \rangle + 1 = \langle x, x \rangle - 1$ so $1 = -1$. Thus σ maps onto $S^n - N$. Now σ is a differentiable function of x. Suppose $\sigma(x) = u$, where u lies in the southern hemisphere. Let us calculate $d\sigma_x(w(x)) = W(u)$. Then $d\sigma_x(w(x)) = \frac{d\sigma}{dt}(x + tw(x))|_{t=0}$, so W is then a nonzero, continuous tangential vector field on the southern hemisphere of S^n. At every point $u = \sigma(u)$ on the equator, $w(u) = u$, points directly outward and $W(u)$ points due north. Similarly using stereographic projection from the south pole on the nonzero vector field $-w(x)$ gives rise to a nonzero, continuous tangential vector field $W(x)$ on the northern hemisphere which also points due north on the equator. Putting these together gives a smooth tangential vector field on the sphere which is certainly nonzero except perhaps at the equator. However, at the equator the vectors add up because they both point north. Hence W is actually a nonzero continuous tangential vector field on S^n. Since n is even this contradicts Corollary 1.2.4. $\square$

The following corollary which is more or less equivalent to the Brouwer fixed point theorem is called *Invariance of Domain*. (S^{n-1} is the boundary of B^n.)

Corollary 1.3.1. *There is no continuous map $g : B^n \to S^{n-1}$ leaving the boundary pointwise fixed.*

Proof. Indeed suppose g is such a map. Let i be the injection of S^{n-1} into B^n and ϕ the map of B^n onto itself sending each vector to its negative. Then ϕ preserves the boundary and the composition $\phi \cdot i \cdot g$ is a continuous map of B^n to itself having no fixed points, a contradiction.

$\square$

1.3.1   A Counter Example

Here we show that, without additional hypotheses, the Brouwer theorem fails even in the case of a separable Hilbert space, H. Namely, that a homeomorphism of the closed unit ball B of H need not have any fixed points (see Kakutani, [63]). As the reader will see extensions of the Brouwer theorem do hold for infinite dimensional spaces when the convex set C in question is *compact (and convex)*. Of course, B cannot be compact unless the space is finite dimensional. (See [90], Theorem 11.2.6, p. 239, where this is proved for any normed linear space.)

Let $(e_n)_{n \in \mathbb{Z}}$ be an orthonormal basis of H. We will define a transformation $T : H \to H$ as follows: Take for T the shift operator, $T(e_n) = e_{n+1}$, and extend linearly and continuously, i.e. if $x = \sum_{n \in \mathbb{Z}} a_n e_n$, then

$$T(x) = \sum a_n e_{n+1}.$$

One sees easily that T is a homeomorphism sending B onto itself preserving the boundary. Now define the map

$$f : H \to H, \quad \text{such that} \quad f(x) := \frac{1}{2}(1 - \| x \|)e_0 + T(x).$$

We will show f is a homeomorphism. First, f is continuous since T is continuous. Also, f is one-to-one. Indeed, for $x \neq 0$ we have

$$f(x) = (1 - \| x \|)\frac{e_0}{2} + \| x \| T\left(\frac{x}{\| x \|}\right).$$

Hence, $f(x)$ divides the line segment joining $\frac{1}{2}e_0$ and $T\left(\frac{x}{\|x\|}\right)$ in the same proportion as the point x divides the segment joining 0 and $\frac{x}{\|x\|}$. This implies if $x \neq y$, then $f(x) \neq f(y)$. Indeed, if x and y are linearly dependent, $f(x)$ and $f(y)$ are different points on the same line segment, and if x and y are linearly independent,

$$T\left(\frac{x}{\|x\|}\right) \text{ and } T\left(\frac{y}{\|y\|}\right)$$

are linearly independent, thus distinct. Therefore $f(x)$ and $f(y)$ are on different line segments and therefore are also distinct.

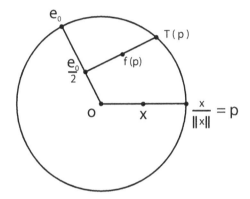

To see that f^{-1} is continuous first observe $\|T(x)\| = \|x\|$. Now from the properties of a norm,

$$\big|\, \|x\| - \|y\| \,\big| \le \|x - y\|. \tag{1.1}$$

Therefore,

$$\|f(x) - f(y)\| = \|\frac{1}{2}(-\|x\| + \|y\|)e_0 + T(x - y)\|$$

$$\ge \big|\, \|T(x-y)\| - \|\frac{1}{2}(\|x\| - \|y\|)e_0\| \,\big|$$

$$\ge \|x - y\| - \frac{1}{2}\big|\, \|x\| - \|y\| \,\big|.$$

Using equation (1.1) we get

$$\|f(x) - f(y)\| \ge \frac{1}{2}\|x - y\|,$$

which shows that f^{-1} is continuous.

We show $f|B$ has no fixed point by contradiction. Suppose $f(x_0) = x_0$. Then,

$$x_0 = f(x_0) = \frac{1}{2}(1- \parallel x_0 \parallel)e_0 + T(x_0).$$

Thus

$$x_0 - T(x_0) = \frac{1}{2}(1- \parallel x_0 \parallel)e_0. \tag{1.2}$$

If $x_0 = 0$, then since $e_0 \neq 0$, $\frac{1}{2}(1- \parallel x_0 \parallel)e_0 = 0$ implies $\parallel x_0 \parallel = 1$, a contradiction. If $\parallel x_0 \parallel = 1$, then $T(x_0) = x_0$. But T has no fixed point on the sphere, since this would imply (e_n) is not an orthonormal basis. Hence

$$0 < \parallel x_0 \parallel < 1.$$

Since (e_n) is an orthonormal basis

$$x_0 = \sum_{n=-\infty}^{+\infty} a_n e_n,$$

where

$$\sum_{n=-\infty}^{+\infty} \mid a_n \mid^2 \leq \parallel x_0 \parallel^2 < 1.$$

But

$$x_0 - T(x_0) = \sum_{n=-\infty}^{+\infty} (a_n - a_{n-1})e_n. \tag{1.3}$$

Since (e_n) is an orthonormal basis , using (1.2) and (1.3), we get

$$a_0 - a_{-1} = \frac{1}{2}(1- \parallel x \parallel) > 0$$

and

$$a_n = a_{n-1}, \quad \text{for } n \neq 0.$$

Thus

$$a_{-3} = a_{-1} = a_{-1} < a_0 = a_1 = a_2 = \cdots$$

which is impossible since

$$\sum_{n=-\infty}^{+\infty} \mid a_n \mid^2 < +\infty.$$

1.3.2   Applications of the Brouwer Theorem

In [75] (see also [76]) it is proved that the adjoint group G, of all *classical*, noncompact, rank 1, simple Lie groups have surjective exponential map. In that proof use is made of the fact that G is the connected group of isometries of a noncompact symmetric space G/K of nonpositive curvature, which is diffeomorphic to the interior of the closed unit ball, B^n. Here K is a maximal compact subgroup of G. Since G has rank 1, each isometry g of G extends continuously to the boundary of G/K. Thus G acts on B^n. Because G is arcwise connected each isometry g is homotopic to the identity. Hence, by the Brouwer fixed point theorem, each g has a fixed point p in B^n. If p lies in the interior of B^n, that is in G/K, since G acts transitively with K the isotropy group of 0, g lies in some conjugate of K (g is an elliptic isometry). Because K is compact and connected, g lies on a 1-parameter subgroup of this conjugate of K (see Corollary 3.4.6) and therefore of G. Otherwise, there is a g-fixed point p on the *boundary* and we are in the case of either a hyperbolic or parabolic isometry, each of which must be analyzed. In the latter case p is unique and in the former there are at least two fixed points on the boundary. The critical case being the parabolic where the Brouwer theorem is again used, but now on a set which is merely homeomorphic with B^n, rather than B^n itself.

   An interesting and recent application of the Brouwer theorem (in the form of Invariance of Domain) due to T. Tao [95] is the following variant of the Hilbert Fifth Problem:

Corollary 1.3.2. *A locally Euclidean topological group is a Lie group.*

1.3.3   The Perron-Frobenius Theorem

Our final application of the Brouwer theorem is the following result of Perron-Frobenius.

Let $A = (a_{ij})$ be an $n \times n$ matrix. We will say that A is *nonnegative* (resp. *positive*), and write $A \geq 0$ (resp. $A > 0$), if $a_{ij} \geq 0$ (resp. $a_{ij} > 0$) for all i, j. Since a vector can be regarded as a 1-row matrix, we can also speak of *nonnegative* (resp. *positive*) *vectors*.

As is customary, we denote by $\mathrm{Spec}(A)$, the *spectrum* of A. That is, the set of all its eigenvalues. The largest modulus among these eigenvalues, is called the *spectral radius* of A, denoted by $r(A)$.

The theorem, proved by Oskar Perron (1907)[1] and Georg Frobenius (1912)[2], asserts that a real, positive square matrix has a unique largest real eigenvalue and the corresponding eigenvector is strictly positive.

This theorem has important applications to probability theory, dynamical systems, economics (Leontief's input-output model), demography (Leslie's population age distribution model), the mathematical background of internet search engines such as Google and even to the ranking of professional tennis players.

Write $v \geq w$ to indicate that no coordinate of the vector v is smaller than the corresponding one of w. If every entry of a matrix A is nonnegative, then $v \geq w$ implies $Av \geq Aw$ since the difference of the ith

---

[1]Oskar Perron (1880-1975) He was a professor at the University of Heidelberg (1914-1922) and at the University of Munich (1922-1951), and although formally retired in 1951, he continued to teach courses at Munich until 1960. Even when he ended his teaching at the age of 80 he still continued a vigorous research program, publishing 18 papers between 1964 and 1973 and 218 (!) papers in total, making numerous contributions to differential and partial differential equations, including the Perron method for solving the Dirichlet problem for elliptic partial differential equations. He also worked on matrices and other topics in algebra, continued fractions, geometry and number theory.

[2]Ferdinand Georg Frobenius (1849-1917) was a student of Weierstrass at the university of Berlin (1870). He is considered, together with Cayley and Sylvester, as one of the principal developers of matrix theory. To him we owe the first rigorous proof of the Cayley-Hamilton Theorem. He was also the mentor and a collaborator of Issai Schur with whom he worked on group theory.

entry is

$$\sum_{j=1}^{n} a_{ij}(v_j - w_j)$$

and all parts of this expression are assumed nonnegative. The same expression shows, if every entry of A is positive, and some entry of v is strictly greater that the corresponding entry of w, then every entry of Av is strictly greater than the corresponding entry of Aw. When $v \geq 0$, the same is true of Av and if $A > 0$, $Av > 0$.

Lemma 1.3.3. *If $A > 0$ and v is an eigenvector of A with $v \geq 0$, then $v > 0$.*

Proof. Since $Av = \lambda v$, Av has zero entries in the same locations as v. Hence, none of the entries of v can be zero.   □

We observe since eigenvectors of A are also eigenvectors of all powers of A, if some power of A is positive, any eigenvector of A with nonnegative entries must actually have strictly positive entries.

Definition 1.3.4. A $n \times n$ matrix A is called a *Markov matrix* (or *stochastic matrix*) if all entries are nonnegative and the sum of the elements of each column is 1.

Proposition 1.3.5. *A Markov matrix A has always the eigenvalue 1, and all other eigenvalues are in absolute value smaller than or equal to 1.*

Proof. For the transpose matrix, A^t, the sum of the elements of each row is 1. Therefore A^t has the eigenvector $(1, ..., 1, ..., 1)^t$. Now because A and A^t have the same determinant, they have the same characteristic polynomial and hence the same eigenvalues. Since A^t has eigenvalue 1, so does A. Now assume that v is an eigenvector with an eigenvalue $|\lambda| > 1$. Then $A^n v = |\lambda|^n v$ has exponentially growing length as $n \to \infty$. Thus for large n, some coefficient of the matrix A^n is larger than 1 and because A^n is a stochastic matrix all its entries are ≤ 1, a contradiction.   □

Theorem 1.3.6. (Perron-Frobenius Theorem[3].) *Let* $A = (a_{ij})$ *be a real, strictly positive, $n \times n$ matrix. Then:*

1. *A has a positive eigenvalue λ with $r(A) = \lambda$.*

2. *λ is a simple eigenvalue.*

3. *The corresponding eigenvector v is also strictly positive.*

4. *A has no other eigenvector with all nonnegative entries.*

5. *An estimate of λ is given by:*

$$\min_i \left(\sum_j a_{ij} \right) \le \lambda \le \max_i \left(\sum_j a_{ij} \right).$$

Proof. Let S^n be the unit sphere centered at the origin in $\mathbb{R}^n$ and set

$$S := \{ v = (v_1, ..., v_n) \mid \|v\| = 1, \ v_i \ge 0, \ \text{for all } i = 1, 2, ..., n \}.$$

One sees easily that S is homeomorphic to the closed unit ball B^{n-1} in $\mathbb{R}^{n-1}$. Define the map $f : S \to S$ by

$$f(v) := \frac{Av}{\|Av\|}.$$

Since f is continuous Brouwer's Fixed Point Theorem tells us it must have a fixed point, $v_0 = (v_{0,1}, ..., v_{0,n})$, i.e.

$$\frac{Av_0}{\|Av_0\|} = v_0.$$

---

[3]The first proof of this theorem, for positive matrices was the analytic one given by Perron [84] in 1907. Shortly afterwards, Frobenius, in a series of papers [43] and [44], extended the result for nonnegative matrices and proved it purely algebraically. Variants of the theorem and its proof were given by Alexandroff and Hopf [3] pp. 480-481, in 1935 and Debreu and Herstein [30] in 1953. In 1956 in [88] Samelson proved the existence of exactly one positive eigenvector by defining a metric in $\text{Int}(\mathbb{R}^+)$, the interior of $\mathbb{R}^+$, and then showing A retracts onto $\text{Int}(\mathbb{R}^+)$. Different constructions were used by Wieland [103] and Brauer [17].

Set $\lambda = \|Av_0\|$ so that $Av_0 = \lambda v_0$, and obviously $\lambda > 0$. By assumption all components of v_0 are nonnegative and $A > 0$. So (even if v_0 has some zero component) Av_0 must be > 0 (see Lemma 1.3.3). Hence, since $\lambda > 0$, the corresponding eigenvector v_0 is positive.

We now show λ has no other eigenvector by contradiction. Indeed, suppose there is another eigenvector x, not lying on the line determined by v_0. Since A and λ are both > 0, the equation, $Ax = \lambda x$ implies, as above, that $x > 0$. Set

$$t = \min_i \frac{v_{0,i}}{x_i} = \frac{v_{0,k}}{x_k},$$

for some k and consider the vector $y = v_0 - tx$. Then

$$Ay = A(v_0 - tx) = Av_0 - tAx = \lambda v_0 - t\lambda x = \lambda(v_0 - tx),$$

i.e. y is also an eigenvector of A associated with the same eigenvalue λ. But

$$y_k = v_{0,k} - tx_k = v_{0,k} - \frac{v_{0,k}}{x_k}x_k = 0,$$

which contradicts the fact that y is an eigenvector of A with eigenvalue λ since v_0 and x are linearly independent.

Considering A^t, which has the same eigenvalues as A and applying the above, there exists a strictly positive w such that $A^t w = \lambda w$. Now, suppose that z is a nonnegative eigenvector of A with eigenvalue $\lambda_1 \neq \lambda$. Then

$$\lambda_1 \langle w, z \rangle = \langle w, \lambda_1 z \rangle = \langle w, Az \rangle = \langle A^t w, z \rangle = \lambda \langle w, z \rangle.$$

Since $\lambda \neq \lambda_1$, $\langle w, z \rangle = 0$, which is impossible since z has all nonnegative entries.

Let $A = (a_{ij})$, $a_{ij} > 0$ (we may even assume $a_{ij} \geq 0$ and none of the column vectors is a zero vector). A first estimate of the value of the eigenvalue λ is

$$\lambda = \|Av_0\| \leq \|A\| \cdot \|v_0\| = \|A\|,$$

where $\|A\|$ is the operator norm coming from the Euclidean (Hilbert-Schmidt) norm on $\mathbb{R}^n$. This is the square root of the maximum of eigenvalues of the matrix AA^t, which does not give us any more numerical information. To find better estimates for λ, we try other norms such as the L_1-norm

$$\|v\|_1 = \sum_i |v_i|,$$

in the definition of S. Then

$$\lambda = \|Av_0\|_1 \leq \|A\|_1 \cdot \|v_0\|_1 = \|A\|_1,$$

where now $\|A\|_1$ is the operator norm with respect to the L_1-norm on $\mathbb{R}^n$, that is the maximum of the L_1-norms of the column vectors of A. To understand the situation, we note that for $\|v_0\| = 1$, the vector Av_0 is a convex linear combination K of the column vectors of A. Hence $\|Av_0\|_1$, which is the distance from the origin, is between $\min_{w \in K} \|w\|$ and $\max_{w \in K} \|w\|$. The minimum and the maximum occur at the vertex points (extreme points) since K is a convex set with flat faces in a hyperplane. Since the vertices of K are given by the column vectors of A, we see that the value $\|Av_0\|_1$ is between the minimum and the maximum of the L_1-norms of the column vectors. In other words,

$$\min_i \left(\sum_j a_{ij} \right) \leq \lambda \leq \max_i \left(\sum_j a_{ij} \right).$$

$\square$

It is important to remark that for a Markov matrix, A, the proof of Theorem 1.3.6 is much simpler. This will become *significant* in applications to computers in the next section. From Proposition 1.3.5 we know that A has 1 as an eigenvalue. Now, define (as above) the set

$$S := \{v = (v_1, ..., v_n) \mid \|v\|_1 = 1, \ v_i \geq 0, \ \text{for all } i = 1, 2, ..., n\}.$$

Then, for any $v \in S$ we get

$$\|Av\|_1 = \sum_i (Av)_i = \sum_i \left(\sum_j a_{ij} v_j \right) = \sum_j v_j \left(\sum_i a_{ij} \right)$$

$$= \sum_j v_j = \|v\|_1 = 1.$$

For the continuous map $f : S \to S$, where here $f(v) := Av$, Brouwer's Fixed Point Theorem tells us there is a v_0 such that $f(v_0) = Av_0 = v_0$ and with all components positive.

1.3.4 Google; A Billion Dollar Fixed Point Theorem

Imagine a huge library containing billions of documents with different formats, no centralized organization and no librarians. Even worse, anyone can add a document at any time without informing anyone else. Under these circumstances how can one hope to quickly find a specific document in this library? Actually, something similar happens whenever one visits the World Wide Web and posed in this way, the problem seems without solution. But this is exactly what happens continuously on the Internet. Using one of the search engines (such as Google, Yahoo, Altavista, etc) one finds the sought after document in some seconds. Back in the 1990's when the first search engines were designed, they used text based systems to decide which web pages were the most relevant to the query. Today's search engines work differently. Simply put they do the following three basic things:

1. They surf the web and locate all web pages with public access.

2. They index the data above, in order to make possible an efficient search for relevant keywords or phrases.

3. They rate, on a probabilistic basis, the importance of each page in the database, so that when a user does a search and the subset of pages in the database with the desired information has been found, the more important pages are presented earlier. It is this step which we will explain.

Google ranks the importance of a webpage using a system named PageRank. This algorithm was developed at Stanford University by Larry Page (hence the name PageRank) and Sergei Brin in 1996, as part of a research project for a new kind of search engine. This was described for the first time in [19]. Shortly afterwards, they founded Google Inc. Brin had the idea that information on the web could be ordered in a hierarchy by "link popularity": a page is ranked higher than another if there are more links to it. That is, if a webpage is linked to another one, this is interpreted as though the first one votes for the second one. The more votes a page obtains, the more important the page is. Moreover, the importance of the page that is casting the vote determines how important the vote is.

The first Google index in 1998 had 26 million pages, and by 2000 the Google index reached the one billion mark. Currently we do not know how big the web is. As engineers at Google say, "time ago", their system which process links on the web hit a milestone: 1 trillion unique URLs on the web at once! These calculations are redone once a month.

To give some idea of how they calculate the rating of a webpage, suppose the webpage A has pages $T_1, \ldots, T_n$ pointing to it (i.e. have citations). Let $\sharp(A)$ be the number of links emanating from A.

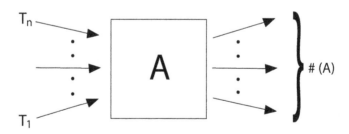

Google did not want a webpage to have a high rank solely because it has a lot of (perhaps contrived) links with other pages. Suppose the page T_j has a total of n_j (outgoing) links to other pages, of which only one is linked to the page A. Then it should increase the rate of A by merely $PR(T_j)/n_j$, where $PR(T_j)$ is the PageRank (that is the importance of T_j). Therefore assume the internet has n webpages, A_k, $k = 1, \ldots, n$

and let N_k be the number of outgoing links from A_k. Then[4],

$$PR(A_k) = \sum_{j \neq k} \frac{PR(A_j)}{N_j}.$$

Here the various $PR(A_j)$ do not influence the value of $PR(A_k)$ uniformly, since the more the page A_j has outgoing links, the less the page A_k will benefit from a link to it from A_j. We also remark first that the different PageRanks form a probability distribution over the webpages, so

$$\sum \text{all webpage's PageRanks} = 1$$

and secondly that $PR(A_k)$ corresponds to the principal eigenvector of the normalized link matrix of the web. Therefore it can be calculated by the above algorithm.

To see how all this works, let us suppose that the web contains only 4 pages, 1, 2, 3 and 4, as in the following figure.

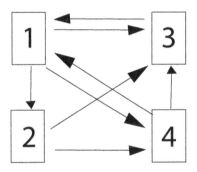

(An arrow from page 1 to 2 shows a link from one webpage to the other.)

Let $PR(1) = x_1$, $PR(2) = x_2$, $PR(3) = x_3$ and $PR(4) = x_4$. Applying what we described above we get

$$x_1 = \frac{x_3}{1} + \frac{x_4}{2}, \quad x_2 = \frac{x_1}{3}, \quad x_3 = \frac{x_1}{3} + \frac{x_2}{2} + \frac{x_4}{2}, \quad x_4 = \frac{x_1}{3} + \frac{x_2}{1}.$$

---

[4]Actually, the formula they use is slightly more complicated in so far as it also involves a parameter $0 \leq d \leq 1$ called *the damping factor*. Then, $PR(A_k)$ is given by $PR(A_k) = (1 - d) + d \sum_{j \neq k} \frac{PR(A_j)}{N_j}$.

Therefore, if we consider the vector $x = (x_1, x_2, x_3, x_4)^t$ we have the system $Ax = x$ with

$$\begin{pmatrix} 0 & 0 & 1 & \frac{1}{2} \\ \frac{1}{3} & 0 & 0 & 0 \\ \frac{1}{3} & \frac{1}{2} & 0 & \frac{1}{2} \\ \frac{1}{3} & \frac{1}{2} & 0 & 0 \end{pmatrix} \cdot \begin{pmatrix} x_1 \\ x_2 \\ x_3 \\ x_4 \end{pmatrix} = \begin{pmatrix} x_1 \\ x_2 \\ x_3 \\ x_4 \end{pmatrix}.$$

Here A has eigenvalue 1 and its *unique* corresponding eigenvector from Perron's theorem. This is $x = (12, 4, 9, 6)$. Normalizing, in order to have a stochastic vector we get $x = \frac{1}{31}(12, 4, 9, 6)$. This ranking makes page 1 the most important webpage, rather than page 3 as one might imagine since it has the most links. This is because page 3, which obviously is an important page, gives its vote only to page 1.

1.4   Fixed Point Theorems for Groups of Affine Maps of $\mathbb{R}^n$

Here we deal, in a primitive form, with some classical fixed point theorems involving groups of affine motions of Euclidean space. In Chapter 2 these will generalize to infinite dimensional spaces as well as Amenable groups and in Chapter 4 to Hadamard manifolds.

The following simple result concerning compact groups will generalize to Élie Cartan's fixed point theorem (see Theorem 4.0.1).

Later we shall also see that the action of G by isometries will of necessity be affine.

Theorem 1.4.1. *Let G be a compact group of isometries of $V = \mathbb{R}^n$ (in the Euclidean metric). Then V has a common G-fixed point.*

Proof. Evidently for fixed $v \in V$ the map $g \mapsto g \cdot v$ is continuous. Let dg be normalized Haar measure on G. Let $v \in V$ and consider $\int_G g \cdot v \, dg$. Using vector valued integration the result is a vector $\bar{v} \in V$, where $\bar{v}$ depends on v (v being fixed once and for all). By left invariance of the measure for $g_1 \in G$,

$$g_1 \cdot \bar{v} = g_1 \cdot \int_G g \cdot v \, dg = \int_G g_1 g \cdot v \, dg = \int_G g \cdot v \, dg.$$

Thus $\bar{v}$ is G-fixed. $\square$

For the reader unfamiliar with Haar measure on compact groups, this is a nontrivial, positive, finite, translation invariant measure on the group. Later, using a fixed point theorem Corollary 2.3.5 we shall prove there is such a measure. Thus, the existence of Haar measure on a compact group yields a fixed point theorem and a more penetrating fixed point theorem will have as a corollary the existence of Haar measure on a compact group. Together with Proposition 1.4.5 below it also shows that compact groups of isometries of Euclidean space are always subgroups of the orthogonal group.

1.4.1   Affine Maps and Actions

Definition 1.4.2. Let V be a real vector space (not necessarily of finite dimension). An operator $A : V \to V$ is called *affine* if $A(v) = T(v) + v_0$ for some $v_0 \in V$, all $v \in V$, where T is a linear operator on V.

A well known characterization of affine operators is the following:

Proposition 1.4.3. *Let* $A : V \to V$ *be an operator. Then* A *is affine if and only if for any* $t \in \mathbb{R}$ *and* v *and* $v' \in V$, $A(tv + (1 - t)v') = tA(v) + (1 - t)A(v')$. *Equivalently, an operator* A *on* V *is affine if and only if it sends convex sets to convex sets.*

Proof. Suppose $A(v) = T(v) + v_0$ is an affine operator. Then

$$\begin{aligned} A(tv + (1 - t)v') &= tT(v) + (1 - t)T(v') + tv_0 + (1 - t)v_0 \\ &= t(T(v) + v_0) + (1 - t)(T(v') + v_0) \\ &= tA(v) + (1 - t)A(v'). \end{aligned}$$

Now suppose A satisfies this condition. If A is to be affine, then v_0 must be $A(0)$. So the only issue is whether $v \mapsto A(v) - A(0)$ is linear. That is,

$$A(v + v') - A(0) = A(v) - A(0) + A(v') - A(0)$$

and

$$A(cv) - A(0) = c(A(v) - A(0)).$$

But by taking $t = \frac{1}{2}$ in the condition in the hypothesis one sees easily that the first of these equations is implied by the second. To prove the second, note that it says, $A(cv) - cA(v) = (1 - c)A(0)$. That is, $A\big(cv + (1 - c)A(0)\big) = cA(v) + (1 - c)A(0)$, which is true. We leave the second statement as an exercise to the reader.   □

Corollary 1.4.4. *A compact group of isometries of $\mathbb{R}^n$ consists of affine transformations.*

Proof. By composing with a translation we can consider the fixed point found above to be a new origin. Since translations are also isometries the new $g \in G$ are again isometries which now preserve the origin and so consists of orthogonal linear transformations. Hence after translating back to the "real origin", we end up with a compact group of isometries which are affine transformations.   □

Proposition 1.4.5. *An isometry of $V = \mathbb{R}^n$ which preserves the origin is a linear transformation. In particular, since a compact group of isometries must fix a point which can be chosen as the origin, any compact group of isometries is a subgroup of the orthogonal group, $O(n, \mathbb{R})$.*

The reader will notice that, as it stands, in the proof below the first statement holds in any Hilbert space.

Proof. For $v, w \in V$ by hypothesis, $\| v - w \|^2 = \| gv - gw \|^2$. Hence

$$\| v \|^2 + \| w \|^2 - 2\langle v, w \rangle = \| gv \|^2 + \| gw \|^2 - 2\langle gv, gw \rangle.$$

But $\| v \|^2 = \| gv \|^2$ and $\| w \|^2 = \| gw \|^2$. Hence $\langle v, w \rangle = \langle gv, gw \rangle$, for any $v, w \in V$. Also,

$$\| g(v + w) \|^2 = \| v + w \|^2 = \| v \|^2 + \| w \|^2 + 2\langle v, w \rangle.$$

On the other hand,

$$\| gv + gw \|^2 = \| gv \|^2 + \| gw \|^2 + 2\langle gv, gw \rangle,$$

and since $\| gv \|^2 = \| v \|^2$, $\| gw \|^2 = \| w \|^2$ and $\langle v, w \rangle = \langle gv, gw \rangle$, we see,

$$\| gv + gw \|^2 = \| g(v + w) \|^2,$$

for any $v, w \in V$.

Finally, we calculate the distance between them. Its square is

$$\| gv + gw - g(v + w) \|^2,$$

which is

$$\| gv + gw \|^2 + \| g(v + w) \|^2 - 2\langle gv + gw, g(v + w) \rangle$$

$$= 2 \| g(v + w) \|^2 - 2(\langle gv, g(v + w) \rangle + \langle gw, g(v + w) \rangle).$$

But this is just,

$$2 \| v + w \|^2 - 2(\langle v, v + w \rangle + \langle w, v + w \rangle) = 0.$$

Thus $g(v + w) = g(v) + g(w)$. From this and the fact that $g(0) = 0$ we see that $g(nv) = ng(v)$ for every integer n. Hence for $n \neq 0$, $g(\frac{1}{n}v) = \frac{1}{n}g(v)$. Therefore also $g(rv) = rg(v)$ for every rational and finally by continuity. $g(tv) = tg(v)$ for every real t. $\square$

We now define an affine action. This will be a fundamental hypothesis of the various fixed point theorems in the next chapter involving groups of operators. The reader should notice that since we have a group, all the affine operators are invertible. Hence all the associated linear operators are also invertible.

Definition 1.4.6. Let V be a real vector space (not necessarily of finite dimension). An action $A : G \times V \to V$ of a group G on V is called *affine* if each of the operators A_g is affine that is, for $g \in G$, $A_g(v) = T_g(v) + w_g$, where each T_g is invertible linear operator and for $g, h \in G$, $A_{gh} = A_g A_h$.

Finally, a continuous affine action is the following.

Definition 1.4.7. Let V be a real locally convex topological vector space and $A : G \times V \to V$ be an affine action. We say A is a *continuous action* if $(g, v) \mapsto A_g(v)$ is a jointly continuous map.

Now given a continuous linear representation $\rho : G \to \mathrm{GL}(V)$, we shall call a continuous function $\varphi : G \to V$ a *1-cocycle* for ρ if for all g_1, $g_2 \in G$ we have

$$\varphi(g_1 g_2) = \rho(g_1)(\varphi(g_2) + \varphi(g_1)).$$

In one sense affine actions reduce to linear ones.

Lemma 1.4.8. *We summarize the salient properties of affine actions.*

1. *$g \mapsto T_g$ is a linear action of G on V.*

2. *$g \mapsto w_g$ is a 1-cocycle with respect to this action.*

3. *If $W = \{(v, t) : v \in V, t \in \mathbb{R}\}$, then there is a linear action ψ of G on W such that the hyperplane $V^* = \{(v, 1) : v \in V\}$ is ψ-invariant and the map $\gamma : v \mapsto (v, 1)$ is a G-equivariant homeomorphism of V with V^*.*

4. *Conversely, let φ be a 1-cocycle with respect to a continuous representation $\rho : G \to \mathrm{GL}(V)$. Then $\varphi(G)$ is a G-orbit under a continuous affine action on V.*

Proof. Now $A(g_1 g_2)(v) = A(g_1) \cdot A(g_2)(v) = A(g_1)(T_{g_2}(v) + w(g_2)) = w(g_1) + T_{g_1} T_{g_2}(v) + T_{g_1}(w(g_2))$. On the other hand this is, $T_{g_1 g_2}(v) + w(g_1 g_2)$, so

$$T_{g_1 g_2}(v) + w(g_1 g_2) = w(g_1) + T_{g_1}(w(g_1)) + T_{g_1}(T_{g_2}(v)). \qquad (1.4)$$

Taking $v = 0$ in (1.4), we get $w(g_1 g_2) = w(g_1) + T_{g_1}(w(g_2))$. Substituting into the equation (1.4) gives $T_{g_1 g_2} = T_{g_1} T_{g_2}$. Next define $\psi_g(v, t) = (T_g(v) + t w_g, t)$ for $g \in G$. Each ψ_g is clearly linear. In fact, ψ_g is a linear operator on W whose "matrix" is

$$\begin{pmatrix} T_g & w(g) \\ 0 & 1 \end{pmatrix}.$$

A direct calculation using the first two statements of the lemma and the fact that each T_g is linear shows that ψ is an action. Since for $v \in V$, $\psi_g(\gamma(v)) = \psi_g(v, 1) = (T_g(v) + w_g, 1)$ while $\gamma A_g(v) = (A_g(v), 1)$ we see V^* is $\psi(G)$ invariant and γ is a G-equivariant homeomorphism. To prove the last item, for $g \in G$, let $A_g : V \to V$ be defined by $A_g(v) = \rho_g(v) + \varphi(g)$. Then each A_g is an affine map on V, and we have a continuous affine action since $A_g(A_h)(v) = \rho_g(\rho_h(v) + \varphi(g)) + \varphi(h)$ which is $\rho_g(\rho_h(v) + \varphi(g)) + \rho_g(\varphi(h)) = \rho_{gh}(v) + \varphi(gh) = A_{gh}(v)$. Observe that $A_g(\varphi(h)) = \rho_g(\varphi(h)) + \varphi(g) = \varphi(gh)$, so $\varphi(G) = A_G(\varphi(1))$.   □

Theorem 1.4.9. *If a compact group G acts affinely on $\mathbb{R}^n = V$, then any compact, convex, G-invariant set $S \subseteq \mathbb{R}^n$ with nonvoid interior has a G-fixed point.*

To see this we need a lemma. We also include a proposition which will be useful later.

Lemma 1.4.10. *Let $x_0, x_1, \ldots, x_n$ be points of a convex set S and $\lambda_0, \lambda_1, \ldots, \lambda_n$ be the barycentric coordinates of some point in the convex hull ($\lambda_i \geq 0$ and $\sum_i \lambda_i = 1$). If $f(x) = T(x) + b$, $x \in V$, is an affine map, then $f(\sum_i \lambda_i x_i) = \sum_i \lambda_i f(x_i)$.*

Later we shall see the usefulness of this concept in the infinite dimensional space of measures.

Proof. $f(\sum_i \lambda_i x_i) = T(\sum_i \lambda_i x_i) + b = \sum_i \lambda_i T(x_i) + b = \sum_i \lambda_i (T(x_i) + b) = \sum_i \lambda_i f(x_i)$.   □

Continuing the proof of the existence of a G-fixed point (Theorem 1.4.9), let the *center of gravity*, cent(S), be defined by

$$\text{cent}(S) = \frac{\int_S x d\mu}{\mu(S)},$$

where $d\mu$ is Lebesgue measure restricted to S. Since S has nonvoid interior and μ is regular $\mu(S) > 0$. The Lebesgue measure is only defined up to a positive multiple, but the quotient in the formula above cannot see which multiple one has chosen. The integral in the numerator can be defined coordinate wise, or as the limit of finite barycentric sums $\sum_i \lambda_i x_i$, where $\lambda_i \geq 0$ and $\sum_i \lambda_i = 1$. The first of these shows the integral exists while the second shows it is independent of the choice of coordinates. Since any affine operator f preserves barycentric coordinates by Lemma 1.4.10,

$$f(\text{cent}(S)) = \text{cent} f(S).$$

In particular, if S is G-invariant $g(\text{cent}(S)) = \text{cent}(S)$ for every $g \in G$. Thus, the center of mass of S is a G-fixed point.

The above means there must be some G-orbit containing $n + 1$ affinely independent points. For otherwise we could replace V by a smaller space.

We shall need the following important result of Mazur (see [33], p. 416), but here only in a finite dimensional space, where it holds because evidently the convex hull of C is bounded.

Proposition 1.4.11. *The closed convex hull of a compact set C in a real Banach space V is compact.*

Now take for S the convex hull of this G-orbit. Since G is compact so is S by Proposition 1.4.11. Therefore S and so also V has a G-fixed point. We can now use this to prove the converse of Corollary 1.4.4. Namely,

Corollary 1.4.12. *A compact group G of affine transformations of V consists of isometries (and so $\mathbb{R}^n$ has a G-fixed point by Theorem 1.4.1).*

Proof. Here we work the other way. As we just saw, a compact group G of affine transformations has a G-fixed point. Then taking that point as origin, G acts linearly. Hence by Proposition 1.4.5 isometrically. $\square$

1.4.2   Affine Actions of Non Compact Groups

We now use fixed point theory to give a very simple proof (due to the first author) that $H^1(G, V) = (0)$ for a compact, or connected semisimple group G with coefficients in a *finite dimensional* real vector space

V. The compact case is very well known, see e.g. [73], p. 334, and actually (see [37], Theorem 6.0.3), for a compact group all the higher order cohomology groups must vanish as well, even when V is a Banach space. Things are quite different when G is a noncompact connected semisimple Lie group where there only seem to be proofs of the following two results: Let G be a real, connected, semisimple Lie group acting continuously and linearly on a real Banach space V.

1. If none of the simple components is locally isomorphic to $SO_0(n, 1)$ or $SU(n, 1)$, then $H^1(G, V) = (0)$. (Erven-Kazdan [36], Chapter V).

2. If G is simply connected, then $H^1(G, V) = (0)$ (S. Komy [65]).

For a counter example in the case of $SO_0(n, 1)$ (which works equally well for $SU(n, 1)$) see [37], p. 118, or the original proof in [31]. For the readers convenience we recall the definition of the first cohomology group $H^1(G, V)$.

Definition 1.4.13. $H^1(G, V)$ is defined to be the quotient group $\mathcal{Z}^1/\mathcal{B}^1$, where $\mathcal{Z}^1$ is the space of the crossed homomorphisms (or 1-cocycles)

$$\varphi : G \to V \; : \; \varphi(gh) = \varphi(g) + g\varphi(h),$$

and $\mathcal{B}^1$ consists of those φ (or 1-coboundaries) of the form $\varphi(g) = g.v_0 - v_0$, for some v_0 in V and all g in G.

Based on a geometric observation of Milnor ([69]) using affine actions, we prove $H^1(G, V) = (0)$ ($V = \mathbb{R}^n$) dealing with the compact and semisimple cases simultaneously, we have the following unified result:

Theorem 1.4.14. *Let G be a group all of whose finite dimensional real representations are completely reducible. Then for every finite dimensional representation of G on V, $H^1(G, V) = (0)$.*

In particular,

Corollary 1.4.15. *If G contains a connected semisimple subgroup H with G/H either compact or of finite volume, then all finite dimensional real representations ρ are completely reducible. (Of course, if G is compact, or connected semisimple this is so. Hence, in all these cases $H^1(G,V) = (0)$.)*

Proof. Since H is connected semisimple any continuous representation is completely reducible by H. Weyl's theorem (see e.g. [1], p. 175). Moreover, as is proved in Moskowitz [74] (Theorem 1, or Corollary 2), since G/H is either compact or of finite volume, ρ must be completely reducible. $\square$

Now to prove Theorem 1.4.14 we need the following:

Lemma 1.4.16. *If ρ is completely reducible continuous representation of G by affine transformations of V, then ρ admits a fixed point.*

Proof. Identify the space V with the hyperplane $\mathbb{R}^n \times \{1\}$ in $\mathbb{R}^{n+1}$. Now, any representation of G by affine transformations of $V \times \{1\}$ extends uniquely to a linear representation of G on $\mathbb{R}^{n+1}$. Indeed the map $x \mapsto Ax + b$, $x \in V$ extends to the map

$$\begin{pmatrix} x \\ 1 \end{pmatrix} \to \begin{pmatrix} A & b \\ 0 & 1 \end{pmatrix} \cdot \begin{pmatrix} x \\ 1 \end{pmatrix} = \begin{pmatrix} Ax + b \\ 1 \end{pmatrix}$$

which is linear. Since the linear subspace $\mathbb{R}^n \times \{0\}$ is invariant, by hypothesis, there exists a complementary G-invariant subspace W. Then, the intersection

$$W \cap (\mathbb{R}^n \times \{1\})$$

is a fixed point which is not (0) since (0) is not in this hyperplane. $\square$

Proof of Theorem 1.4.14 Let $\rho : G \to \mathrm{GL}(V)$ be a continuous linear representation of G and φ be a 1-cocycle. Define the affine map,

$$\rho_\varphi : G \to \mathrm{Aff}(V) := G \ltimes GL(V),$$

given by

$$\rho_\varphi(g) : V \to V \quad \text{such that} \quad \rho_\varphi(g)(v) := \rho(g)(v) + \varphi(g).$$

From the cocycle identity this map is a homomorphism. But by Lemma 1.4.16, the affine map ρ_φ has a fixed point. That is, there is a v_0 in V with $\rho_\varphi(g)(v_0) = v_0$, for each $g \in G$. Then $\rho(g)(v_0) + \varphi(g) = v_0$ so that φ is a 1-coboundary and $H^1(G, V) = (0)$.

Chapter 2

Fixed Point Theorems in Analysis

Here we discuss several different types of fixed point theorems. One concerns a single *arbitrary continuous* function acting on a compact convex set C in a locally convex real linear topological vector space V. This is a direct generalization of the Brouwer theorem to infinite dimensional spaces. Another type deals with a *group G* of commuting *affine* mappings also acting on a compact convex subset C of a real linear topological vector space V. Here we get simultaneous G-fixed points. A variant concerns equicontinuous, or distal groups of *affine* mappings, also giving simultaneous G-fixed points. The situation of an abelian or a compact group are susceptible to considerable generalization leading us to amenable groups and yielding a fixed point theorem for such groups acting affinely on arbitrary locally convex linear topological vector spaces (Theorem 2.4.4) where our treatment follows Zimmer, [106]. For all these results, even in the finite dimensional case (the Brouwer theorem) convexity is essential. For suppose C consisted of just two distinct points. Then interchanging these points would surely be a continuous map of this compact set which evidently has no fixed points.

2.1   The Schaüder-Tychonoff Theorem

The Schaüder-Tychonoff theorem is the natural generalization of Brouwer's theorem to the context of infinite dimensional space and as such it affords applications to differential and integral equations which elude the Brouwer theorem.

Theorem 2.1.1. *Let V be a locally convex linear topological vector space, C be a compact convex subset of V and $f : C \to C$ a continuous map. Then f has a fixed point.*

We shall first establish some properties of convex sets which will be needed in the sequel. In this chapter C will always denote a nonempty set.

Lemma 2.1.2. *Convex sets enjoy the following properties:*

1. *The intersection of any family of convex sets is convex.*

2. *C is convex if and only if for any finite subset $x_1, \ldots, x_n$ of C and nonnegative numbers $a_1, \ldots, a_n$ with $\sum a_i = 1$, then $\sum a_i x_i \in C$.*

3. *If C is convex and $a \geq 0$, then aC is convex.*

4. *If C_1 and C_2 are convex, then $C_1 \pm C_2$ is convex.*

5. *Let T be an affine map of V, that is, $T(u) = L(u) + v_0$, for $u \in V$ and v_0 fixed. Then for x and $y \in V$ and $\lambda \in \mathbb{R}$,*

$$\lambda T(x) + (1 - \lambda)T(y) = T(\lambda x + (1 - \lambda)y).$$

6. *Let C be a convex subset and T be an affine map of V. Then $T(C)$ is convex.*

Proof. The first four items are obvious. For the last two, let x and $y \in C$ and $0 \leq \lambda \leq 1$. Then

$$
\begin{aligned}
\lambda T(x) + (1 - \lambda)T(y) &= \lambda(L(x) + v_0) + (1 - \lambda)(L(y) + v_0) \\
&= \lambda L(x) + (1 - \lambda)(L(y)) + v_0 \\
&= L(\lambda x + (1 - \lambda)y) + v_0 = T(\lambda x + (1 - \lambda)y).
\end{aligned}
$$

Taking $0 \leq \lambda \leq 1$, the last statement follow from the previous one since C is convex. $\square$

We shall also require some details concerning Hilbert spaces, particularly the Hilbert cube.

Lemma 2.1.3. *Let V be a Hilbert space, $x \in V$ be fixed and $K \subseteq V$ satisfying $\frac{1}{2}(K + K) \subseteq K$ (in particular, for K convex). Choose any sequence $k_i \in K$ satisfying $\lim_{i \to \infty} \| x - k_i \| = \inf_{k \in K} \| x - k \|$. Then k_i converges.*

Proof. Using the Parallelogram Law, we see that for any integers i and j,

$$\| (x - k_i) + (x - k_j) \|^2 + \| (x - k_i) - (x - k_j) \|^2 = 2 \| x - k_i \|^2 + 2 \| x - k_j \|^2 .$$

Hence,

$$\| k_i - k_j \|^2 = 2 \| x - k_i \|^2 + 2 \| x - k_j \|^2 - 4 \| x - \frac{k_i + k_j}{2} \|^2,$$

and, since $\frac{k_i + k_j}{2} \in K$, for i and j sufficiently large,

$$\| k_i - k_j \|^2 = 2 \| x - k_i \|^2 + 2 \| x - k_j \|^2 - 4 \| x - \frac{k_i + k_j}{2} \|^2 .$$

Since each of the terms on the right tend to zero, so does $\| k_i - k_j \|$. Thus this Cauchy sequence converges. $\square$

Corollary 2.1.4. *Let $K \subseteq C$ be a closed convex set of the Hilbert cube C. Then each point $c \in C$ has a unique nearest point $N(c) \in K$. This gives a function $c \mapsto N(c)$ which is continuous.*

Proof. Let $c \in C$ be fixed. Since K is compact there is a sequence $k_i \in K$ converging to c which minimizes. Suppose some other sequence $k_i' \in K$ converging to c' minimizes $\inf_{k \in K} \| c' - k \|$. Since K is compact and so closed, c and $c' \in K$. By Lemma 2.1.3 the sequence $k_1, k_1', k_2, k_2', \ldots$

converges. But subsequences of it converge to both c and c' so $c = c'$ and gives a function $c \mapsto N(c)$ mapping $C \to K$.

To see that N is continuous let $c_n \to c$. We must show $N(c_n)$ converges to $N(c)$. Since K is compact, $N(c_n)$ has some convergent subsequence $N(c_{n_i})$ converging to $k \in K$. We change notation and call this subsequence c_n so that both $c_n \to c$ and $N(c_n) \to k$. Now

$$d(c_n, N(c_n)) \le d(c_n, N(c)) \le d(c_n, c) + d(c, N(c)).$$

The first inequality because of the minimal property of N and the second by the triangle inequality. Hence $d(c_n, N(c_n)) \le d(c, N(c))$. But since $c_n \to c$ and $N(c_n) \to k$ the continuity of d tells us $d(c_n, N(c_n)) \to d(c, k)$ so that also $d(c, k) \le d(c, N(c))$. By the minimal property this means we have equality and $k = N(c)$ so that N is continuous. $\square$

2.1.1   Proof of the Schaüder-Tychonoff Theorem

We will now prove the Schaüder-Tychonoff theorem in the special case where $V = l_2$ and C is the Hilbert cube. However, before doing so we should check that the Hilbert cube, $C = \{x \in l_2 : \| x_n \| \le \frac{1}{n}\}$, actually fits the bill.

Proposition 2.1.5. *The Hilbert cube, C, is compact and convex.*

Proof. Let $x = (x_1, \dots, x_n, \dots)$ and $y = (y_1, \dots, y_n, \dots)$ be points in C. Then, for $0 \le t \le 1$,

$$tx + (1-t)y = \left(tx_1 + (1-t)y_1, \dots, tx_n + (1-t)y_n, \dots \right).$$

For each n,

$$\| tx_n + (1-t)y_n \| = t \| x_n \| + (1-t) \| y_n \| \le \frac{1}{n}(t + (1-t)) = \frac{1}{n}.$$

So C is convex.

To see C is compact, since l_2 is a complete metric space it is sufficient to show C is totally bounded. Let $\epsilon > 0$. For $x = (x_1, \dots, x_n, \dots) \in C$

choose n sufficiently large so that the tail of the convergent series is small.

$$\sum_{i=n+1}^{\infty} \frac{1}{i^2} < \frac{\epsilon}{2}.$$

Then, since $\| x_i \| \leq \frac{1}{i}$ for all i, it follows that,

$$\sum_{i=n+1}^{\infty} \| x_i \| < \frac{\epsilon}{2}.$$

Let $x_* = (x_1, \ldots, x_n, 0, \ldots)$. Then $d(x, x_*) < \frac{\epsilon}{2}$. Moreover, for this fixed n we are in a finite dimensional Euclidean space $\mathbb{R}^n$. Hence the set of x_* coming from x's in C is closed and bounded and therefore compact. Thus there is a finite $\frac{\epsilon}{2}$ net which covers $C \cap \mathbb{R}^n$. Evidently the corresponding finite ϵ net covers C.   □

We next prove Theorem 2.1.1 in the case that $V = l_2$ and C is the Hilbert cube.

Lemma 2.1.6. *Let C be the Hilbert cube and f a continuous map of C to itself. Then f has a fixed point.*

Proof. Let $f : C \to C$ be a continuous self map of C and $\epsilon > 0$. Then for $c \in C$, $f(c) = (c_1, \ldots, c_n, \ldots)$. For each positive integer n, let $f_n(c) = (c_1, \ldots, c_n, 0, \ldots)$. Then $f_n(C)$ is homeomorphic to $C_n = I \times \frac{1}{2} I \times \ldots \times \frac{1}{n} I$ which in turn is itself homeomorphic to the closed unit ball in $\mathbb{R}^n$. On the other hand $f_n \cdot f(C) \subseteq f_n(C) = C_n$ so that $f_n \cdot f(C_n) \subseteq C_n$ and since f is continuous so is $f_n \cdot f$. By the Brouwer theorem each f_n has a fixed point c_n and all these c_n are in C. Now,

$$d(c_n, f(c_n)) = \| c_n - f(c_n) \| \leq \sqrt{\sum_{i=n+1}^{\infty}} < \epsilon,$$

if n is large. Since $c_n \in C$, which is compact, a subsequence converges to $c \in C$. By continuity $f(c_{n_i})$ converges to $f(c)$. Therefore $d(c, f(c)) < 2\epsilon$.   □

We now extend our result to arbitrary compact convex subsets of the Hilbert cube.

Lemma 2.1.7. *Let C be the Hilbert cube and K be a compact convex subset of C and f a continuous map of K to itself. Then f has a fixed point.*

Proof. Since $N(C) \subseteq K$ we know N stabilizes K as does f. Hence so does $f \cdot N$. But for $c \in K$, $N(c)$ is *the* point of K nearest c. Since c is clearly this point, $c = N(c)$. Now $f \cdot N : C \to K \subseteq C$. As a continuous map of the Hilbert cube it has a fixed point $c \in K$. So $f \cdot N(c) = c$. Hence $f(c) = c$.   □

The key Lemma needed to prove the Schaüder-Tychonoff theorem in general comes next, but requires the following definition.

Definition 2.1.8. Let Λ and $M \subset V^*$. We shall say Λ is *determined* by M if for each $\epsilon > 0$ there is a finite subset F of M, a $\delta > 0$ and a neighborhood of zero

$$N(0, F, \delta) = \bigcap_{\mu \in F} \{x : |\mu(x)| < \delta\}$$

with the property that if p and $q \in C$ and $p - q \in N$, then

$$|\lambda(f(p)) - \lambda(f(q))| < \epsilon,$$

for all $\lambda \in \Lambda$.

Lemma 2.1.9. *Let V be a locally convex linear topological vector space, C be a compact convex subset of V containing at least two points and $f : C \to C$ a continuous map. Then there is a proper closed convex subset C_1 of C which is also f invariant.*

Proof. Consider the weak topology on V. Namely, a net $x_i \to x$ in V if for each fixed $\lambda \in V^*$ (or finite set of λ's), $\lambda(x_i) \to \lambda(x)$ in $\mathbb{R}$. Since the identity map $I : V \to V$ (where on the left we have the given topology and on the right the weak topology) is continuous, so is its restriction

$C \to C$. Because C is compact it is a homeomorphism on C. Moreover I is affine hence the image of C is convex. Thus we may replace the given topology on C by the weak topology.

Suppose Λ is determined by M and $\mu(p) = \mu(q)$ for all $\mu \in F$. Then $|\mu(p - q)| = 0 < \delta$ for all $\mu \in M$. Hence $|\lambda(f(p)) - \lambda(f(q))| < \epsilon$ for all $\lambda \in \Lambda$ and since ϵ is arbitrary this forces $\lambda(f(p)) = \lambda(f(q))$.

We first observe that each $\lambda \in V^*$ is determined by a *countable* set of functionals M_λ. This is because the numerical function $p \mapsto f(\lambda(p))$ for $p \in C$ is continuous and so because C is compact, it is uniformly continuous. Hence for each positive integer n, taking $\epsilon = \frac{1}{n}$ there is a finite set $M_n \subseteq V^*$ and $\delta_n > 0$ so that the neighborhood, $N(0, M_n, \delta_n)$, of 0 has the property that if $p, q \in C$ and $p - q \in N$, then $|\lambda(f(p)) - \lambda(f(q))| < \frac{1}{n}$. Then $\cup_{n=1}^{\infty} M_n$ as a countable union of finite sets is countable. From this it follows that if Λ is a countable set of functionals it is determined by a countable set M_Λ because a countable union of countable sets is still countable. Indeed, we can go somewhat further and show that each $\lambda \in V^*$ can be included in countable set of self determined functionals. For if λ is determined by the countable set M_1 and by the previous statement each functional in M_1 is determined by the countable set M_2 etc. then $M = \lambda \bigcup \cup_{i=1}^{\infty} M_i$ is clearly countable and self determined.

Now suppose C contains two distinct points p and q. Choose a $\lambda \in V^*$ with $\lambda(p) \neq \lambda(q)$ and let $M = \{\lambda_i\}$ be a countable set of self determined functionals containing $\lambda = \lambda_1$. Since C is compact each $\lambda_i(C)$ is bounded. Hence multiplying each λ_i by a suitable positive constant we still have a countable set of self determined functionals containing the new λ_1 and we can arrange that for every i, $|\lambda_i(C)| \leq \frac{1}{i}$. This gives rise to a map L of C into the Hilbert cube whose i^{th} coordinate is given by $c \mapsto \lambda_i(c)$. Because $\lambda_1(p) \neq \lambda_1(q)$, $L(p) \neq L(q)$ and since L is a *linear* and continuous map, $L(C) = C_0$ is a compact convex subset of the Hilbert cube which also contains at least the two distinct points $L(p)$ and $L(q)$.

We can assume L is injective. For if $L(c) = L(d)$ for some c and $d \in C$, that is, $\lambda_i(c) = \lambda_i(d)$ for all $i \geq 1$, since $\{\lambda_i; i \geq 1\}$ is self determined, also $f\lambda_i(c) = f\lambda_i(d)$ for all i. So in any case, fL^{-1} is well defined. Let

$f_0 = LfL^{-1}$, then $f_0(C_0) = LfL^{-1}L(C) = Lf(C) \subseteq L(C) = C_0$ so C_0 is f_0-invariant. We show f_0 is continuous. Let $c_0 \in C_0$ and $0 < \epsilon < 1$. Choose N large enough so that $\sum_{i=N}^{\infty} \frac{1}{i^2} < \epsilon$. Because the λ_i is a set of self determined functionals there is a $\delta > 0$ and a positive integer m so that if $|\lambda_j(p) - \lambda_j(q)| < \delta$ for $j = 1, \ldots, m$, then for $i = 1, \ldots, N$.

$$|\lambda_j(f(p)) - \lambda_j(f(q))| < \sqrt{\frac{\epsilon}{N}}.$$

Now take c so that $\| c - c_0 \| < \delta$. Then

$$\| f_0(c) - f_0(c_0) \|^2 = \| LfL^{-1}(c) - LfL^{-1}(c_0) \|^2$$

$$\leq \sum_{i=1}^{N} |\lambda_i(f(p)) - \lambda_i(f(q))|^2 + 2 \sum_{i=N}^{\infty} \frac{1}{i^2} < 3\epsilon.$$

Thus f_0 is continuous on all of C_0. By Lemma 2.1.7 f_0 has a fixed point $c_0 \in C_0$. But then $fL^{-1}(c_0) \subseteq L^{-1}f_0(c_0) = L^{-1}(c_0)$. Let $C' = L^{-1}(c_0)$. Then $f(C') \subseteq C'$ and C' is a closed and therefore compact subset of C. If both $L(x) = c_0$ and $L(y) = c_0$, then because L is linear $L(tx + (1-t)y) = tc_0 + (1-t)c_0 = c_0$ so C' is convex. Finally, we can take $C' = L^{-1}(c_0)$ to be a proper subset of C because if $C = L^{-1}(c_0)$, then $f(C) = \{c_0\}$ and so c_0 would already be a fixed point of f and so we would already be done. $\square$

Completion of the proof of the Schaüder-Tychonoff theorem:

Proof. Consider the family of all nonempty, closed, convex, G-invariant subsets of C of which C' is a typical member. Order these by $\supseteq$. This is a partially ordered set. Take any chain in this family. Then the intersection of the elements of the chain is G-invariant, closed and convex. Since any finite intersection of the elements of this chain is nonempty, the entire intersection is nonempty as well by the finite intersection property. Thus every chain has a smallest element. By Zorn's lemma there is minimal element among all the nonempty, closed, convex, G-invariant sets $C' \subseteq C$, which we again denote by C. By Lemma 2.1.9 this must consist of a single point. $\square$

2.2 Applications of the Schaüder-Tychonoff Theorem

Here we illustrate the power the Schaüder-Tychonoff theorem. As mentioned earlier, because the space V can be infinite dimensional it can be used to solve both integral and differential equations.

We first turn to integral equations. The reader will notice that in these applications of the Schaüder-Tychonoff theorem use is made of the Ascoli-Arzela theorem. Let X be a compact, Hausdorff space and $C(X)$ denote the complex continuous functions on X with the sup norm. A kernel function $K : X \times X \to \mathbb{C}$ means a jointly continuous function of 2 variables. Of course if K takes only real values, then $C(X)$ will denote the real valued functions on X.

Corollary 2.2.1. *Let X be a compact, Hausdorff space, μ a finite regular measure on X and K be a kernel function. Then the following integral equation always has a solution $f \in C(X)$.*

$$f(x) = \int_X K(x, y)f(y)d\mu(y). \qquad (2.1)$$

Proof. For $f \in C(X)$ let $T(f)$ be defined by the right side of equation (2.1). We first show $T(f)$ is a bounded and therefore continuous function on X. This is because K and f are bounded and μ is finite so that,

$$|T(f)| = \left| \int_X K(x, y)f(y)d\mu(y) \right| \le |K|_X |f|_X \mu(X).$$

Hence T is a bounded linear operator on $C(X)$ with operator norm $|K|_X \mu(X)$. For convenience we choose μ so that $|K|_X \mu(X) = 1$.

Actually, T is a completely continuous operator. That is, it sends bounded sets into *compact* sets. Since X is compact and K is continuous, by uniform continuity given $\epsilon > 0$, we know $|K(x, y) - K(x', y')| < \epsilon$, if (x, y) and (x', y') are "near enough" to one another. Hence for x and x' near enough

$$|T(f)(x) - T(f)(x')| \le \int_X |K(x, y) - K(x', y)||f(y)|d\mu(y) \le \epsilon |f|_X.$$

Thus for a norm bounded set of f's, $\{T(f)\}$ is uniformly equicontinuous on X. It is also uniformly bounded by the above. Hence by Ascoli-Arzela, $T(B)$ has compact closure, where B is the closed unit ball in $C(X)$. Thus T is completely continuous. Since B is convex and T is linear, $\overline{T(B)}$ is also convex. Now the compact convex set $\overline{T(B)}$ lies within B. Hence $T(\overline{T(B)}) \subseteq \overline{T(B)}$ so that $\overline{T(B)}$ is T-stable. The Schaüder-Tychonoff theorem guarantees there is a fixed point $f \in \overline{T(B)} \subseteq B \subseteq C(X)$. That is, f satisfies the integral equation (2.1). □

We now turn to Peano's theorem.

Corollary 2.2.2. (Peano's Theorem.) *Let $f(x,t)$ be a continuous and bounded function on the plane, $\mathbb{R}^2$. Then the initial value problem,*

$$\frac{dx(t)}{dt} = f(x(t),t), \quad x(0) = 0,$$

has a global *C^1 solution. That is, the solution holds for all $t \in \mathbb{R}$.*

Before beginning the proof we remark that a family of C^1 real functions $x(t)$ defined on an interval, whose derivatives are uniformly bounded, say by b, on that interval is equicontinuous. This is because of the Fundamental Theorem of Calculus, $x(t_1) - x(t_2) = \int_{t_1}^{t_2} x'(t)dt$. Hence, $|x(t_1) - x(t_2)| \leq \int_{t_1}^{t_2} |x'(t)|dt \leq b|t_1 - t_2|$.

Proof. Let $|f(x,t)| \leq M$ on $\mathbb{R}^2$ and suppose we have a solution $z(t)$ to the initial value problem on $[0, L]$. Then again by the Fundamental Theorem of Calculus,

$$z(t) = \int_0^L f(\eta, z(\eta))d\eta. \tag{2.2}$$

Conversely, if $z(t)$ satisfies equation (2.2) it is a solution to the initial value problem on $[0, L]$. If we can solve the initial value problem for arbitrary L then letting $L \to \pm\infty$ will yield a solution on all of $\mathbb{R}$.

Now let $x(t)$ be *any* continuous function $[0, L] \to \mathbb{R}$, and let y be defined by $y(t) = \int_0^L f(\eta, x(\eta))d\eta$. As we shall see the map taking $x \mapsto y$, which we denote by F goes from $C^0[0, L] \to C^1[0, L]$.

We first show F is continuous at each point $x_0 \in C^0[0, L]$ (sup norm topology),

$$\| F(x) - F(x_0) \| \leq \int_0^L |f(\eta, x(\eta)) - f(\eta, x_0(\eta))|d\eta.$$

Since f is continuous, given $\epsilon > 0$ there is a $\delta > 0$ so that

$$|f(\eta, x(\eta)) - f(\eta, x_0(\eta))| < \epsilon, \text{ if } |x(\eta) - x_0(\eta)| < \delta.$$

Hence if $\| x - x_0 \| < \delta$ this is so. Therefore we see whenever $\| x - x_0 \| < \delta$ we get $\| F(x) - F(x_0) \| \leq \epsilon L$. Thus F maps from C^0 to itself.

To see that F actually maps from C^0 to C^1, note that

$$F(x)(t) - F(x)(t_0) = \int_{t_0}^t f(\eta, x(\eta))d\eta.$$

By the mean value theorem for integrals, $\frac{F(x)(t) - F(x)(t_0)}{t - t_0} = f(\bar\eta, x(\bar\eta))$, where $\bar\eta$ is between t and t_0. Taking $\lim_{t \to t_0}$ and using the continuity of f we see this is $f(t_0, x(t_0))$ so the limit exists and gives a continuous function. Thus $F(x)$ is $C^1[0, L]$.

Now on C^1 we take for the norm the sum of the C^0 sup norm and the sup norm of the derivative. Thus

$$\| F(x) \|_1 = \| F(x) \|_0 + \| F'(x) \|_0 .$$

Now evidently, $\| F(x) \|_0 \leq LM$ and $\| F'(x) \|_0 \leq M$. Hence, $\| F(x) \|_1 \leq LM + M = (L + 1)M$.

Denote by $S = \{x \in C^0[0, L] : \| x \|_0 \leq LM\}$. Then $F(x) \in S$, for all $x \in C^0[0, L]$ and in particular for all $x \in S$, so F maps S to itself. Let T be the subset of S defined by the additional conditions $x \in C^1[0, L]$ and $\| x \|_1 \leq (L + 1)M$. Then T is clearly closed in $C^1[0, L]$ (but as one would expect, not in $C^0[0, L]$ since $C^1[0, L]$ is dense in $C^0[0, L]$). As we showed above T and so also its closure is F invariant. T is also clearly

convex. Moreover, the closure of T in $C^0[0, L]$ is compact since it is
an equicontinuous family of functions due to the initial Remark (2.2).
Thus the closure of T in $C^0[0, L]$ is both compact and convex and F
stable. By the Schaüder-Tychonoff theorem F has a fixed point in the
closure of T which is clearly a solution to the initial value problem. □

2.3   The Theorems of Hahn, Kakutani and Markov-Kakutani

As opposed to merely a single continuous map, in this section we shall
study groups of maps acting on linear topological vector spaces, but
here in compensation we shall need to assume they are affine. We first
give F. Hahn's elegant proof of his fixed point theorem for groups of
distal actions by affine maps. As a consequence we get Kakutani's fixed
point theorem for equicontinuous groups of affine maps. (Of course, in
particular these results hold for linear actions by distal or equicontinuous
families of maps.) By the theorem of Ascoli-Arzela, Kakutani's theorem
holds when G is a compact group (see [33], p. 266). Finally, we deal with
abelian group actions and the fixed point theorem of Markov-Kakutani.

Definition 2.3.1. Let $G \times C \to C$ be a jointly continuous group action
of G on C where G is a topological group and C is a compact space.
We say this action is *distal* if for x and $y \in C$ and a net $g_i \in G$, if
$\lim g_i(x) = \lim g_i(y)$, then x and y must be equal.

Of course, if C is a metric space with metric d what this means is
for each pair of distinct points x and $y \in C$, $\inf_{g \in G} d(g(x), g(y)) > 0$.

Theorem 2.3.2. *Let V be a real locally convex linear topological vector
space, C a compact, convex set in V and G be a group acting affinely
and jointly continuously on V leaving C stable. If the action on C is
distal, then C has a G-fixed point.*

Proof. By Zorn's lemma exactly as in the proof of Theorem 2.1.1 we
know there is a minimal f-invariant compact convex set C' within C.
Since this is contained in the original C, G also acts distally on it. We

rename C' as C. Of course, if C consists of just a single point we are done. Otherwise, let x and y be distinct points of C and $z = \frac{x+y}{2}$, which by convexity is also in C. Let $\mathcal{O}$ be the G-orbit closure of z in C. Then $\mathcal{O}$ is, of course, closed and G-invariant. Since C is compact and G-invariant $\mathcal{O} \subseteq C$ and so is also compact and since C is convex the convex hull, $\mathcal{O}^c$, of $\mathcal{O}$ is contained in C. By its minimality C is the closed convex hull of $\mathcal{O}$. By the Krein-Milman theorem (see [33], p. 440) as a compact convex set, C is the convex hull of its extreme points. Let p be an extreme point of C, $p \in \mathcal{O}$. Hence there is a net g_i in G so that $\lim g_i(z) = p$.

By compactness of $\mathcal{O}$ there is a subnet (which we again call g_i) with $\lim g_i(x) = a$ and another (also called g_i) with $\lim g_i(y) = b$, where a and $b \in C$. Taking appropriate subnets we can assume these g_i are the same and taking that subnet of the original net we can assume $\lim g_i(z) = p$, $\lim g_i(x) = a$ and $\lim g_i(y) = b$. Since the action is by affine maps we see that,

$$\lim g_i(z) = \lim g_i\left(\frac{x+y}{2}\right) = \lim \frac{g_i(x) + g_i(y)}{2} = \frac{a+b}{2}.$$

Hence $p = \frac{a+b}{2}$. But since p is extreme point of C and a and $b \in C$ it follows that $a = p = b$ and because x and y are distinct and the action is distal, this is a contradiction.   □

Definition 2.3.3. Let $G \times C \to C$ be a jointly continuous group action of G on C where G is a topological group and C is a compact space. We say this action is *equicontinuous* if for any neighborhood U_0 of 0 in V, there is a neighborhood U_1 of 0 so that $G \cdot U_1 \subseteq U_0$.

As a consequence of Theorem 2.3.2 we have,

Theorem 2.3.4. *Let V be a real locally convex linear topological vector space, C a compact, convex set in V and G be a group acting affinely and jointly continuously on V leaving C stable. If G acts equicontinuously on C, then C has a G-fixed point.*

Proof. For the proof we simply show that an equicontinuous action must be distal (whether or not the action is affine). Let x and y be distinct points of C and choose a neighborhood U_0 of 0 in V so that $x - y$ is not in U_0. By equicontinuity of the action there is a neighborhood U_1 of 0 so that $gu - gv \in U_0$ for all $g \in G$ whenever $u - v \in U_1$. Hence for each $g \in G$, $gx - gy$ is not in U_1. For if it were for some g, then for that g, $g^{-1}(gx - gy) = x - y$ would be in U_0, a contradiction. This means that for no net g_i can $\lim g_i x = \lim g_i y$ and so the action is distal. $\qquad\square$

As an application of Theorem 2.3.4 we prove the existence of Haar measure on a compact group. There seems to be no fixed point theorem type argument for the existence of (say left) Haar measure on a locally compact group in general.

Corollary 2.3.5. *Let G be a compact topological group. Then G has a nontrivial positive, regular G-invariant measure.*

Proof. Let $\mathcal{P}_{reg}(G)$ be the regular measures on G and $\mathcal{P}_{reg}(G)^+$ the positive regular measures on G of total mass 1. Let G act on $\mathcal{P}_{reg}(G)$ by left translation: $(g \cdot \mu)(A) = \mu(gA)$, where $g \in G$ and A is a Borel measurable set in G. Then G acts linearly because if μ and ν are regular measures on G, then

$$g \cdot (\mu + \nu)(A) = (\mu + \nu)(gA) = \mu(gA) + \nu(gA) = g \cdot \mu(A) + g \cdot \nu(A),$$

and similarly, $g \cdot c\mu = cg \cdot \mu$. Moreover, if $\mu \in \mathcal{P}_{reg}(G)^+$, then each $g \cdot \mu$ is positive and regular. Also, $(g \cdot \mu)(G) = \mu(gG) = \mu(G) = 1$. Hence $\mathcal{P}_{reg}(G)^+$ is G-invariant. $\mathcal{P}_{reg}(G)^+$ is certainly convex because if $0 \le t \le 1$ and μ and $\nu \in \mathcal{P}_{reg}(G)^+$, then $t\mu + (1 - t)\nu \in \mathcal{P}_{reg}(G)^+$ since it is positive, regular and

$$(t\mu + (1 - t)\nu)(G) = t\mu(G) + (1 - t)\nu(G) = t1 + (1 - t)1 = 1.$$

Finally by the theorem of Alaoglu, $\mathcal{P}_{reg}(G)^+$ is compact in the weak* topology (see [33], p. 424). Since a G-invariant measure is exactly a simultaneous G-fixed point, an application of Theorem 2.3.4 completes the proof. $\qquad\square$

As is well known, a locally compact group G has finite Haar measure if and only if it is compact. Since, except for the application of the

Kakutani fixed point theorem, the proof of Corollary 2.3.5 works for any locally compact group, this shows the hypothesis of equicontinuity is essential for Theorem 2.3.4.

We now come to a fixed point theorem is due independently to Markov and Kakutani. Notice that this result does not require V to be locally convex.

Theorem 2.3.6. *Let V be a real linear topological vector space, C be a compact, convex set in V and G be a commutative group acting affinely on V leaving C stable. Then C has a G-fixed point.*

Proof. For each positive integer n and $g \in G$ let $M_{n,g} = \frac{1}{n}(I + g + \cdots + g^{n-1})$. Each $M_{n,g}$ is an operator on V. Since C is convex and G-invariant, each $M_{n,g}(C) \subseteq C$. Let M^* denote the semigroup generated by all the $M_{n,g}$, that is all finite compositions of these operators. Since G is abelian so is M^* and since each $M_{n,g}$ leaves C invariant so does each $T \in M^*$. We will show $\bigcap_{T \in M^*} T(C)$ is nonempty and that each element of this intersection is a G fixed point.

To see that $\bigcap_{T \in M^*} T(C)$ is nonempty, first observe that each $T(C)$ is compact because each T is continuous and C itself is compact. By the finite intersection property it suffices to show that for any finite set $T_1, \ldots, T_k$ that $\bigcap_{i=1}^{k} T_i(C)$ is nonempty. Set $S = T_1 \cdots \cdots T_k$. By commutativity for each i,

$$S(C) = (T_1 \cdots \cdots T_k)(C) = (T_i \cdot T_1 \cdots T_i \cdot T_k)(C) \subseteq T_i(C).$$

Hence $S(C) \subseteq \bigcap_{i=1}^{k} T_i(C)$ and since of course $S(C)$ is nonempty so is $\bigcap_{i=1}^{k} T_i(C)$.

Now let $y \in \bigcap_{T \in M^*} T(C)$. Then for each n and g we have $y \in M_{n,g}(C)$. That is, $y = \frac{1}{n}(x + g(x) + \ldots + g^{n-1}(x))$ for some $x \in C$. But since G acts by affine operators $g(y) = \frac{1}{n}(g(x) + \ldots + g^{n-1}(x) + g^n(x))$. Hence $g(y) - y = \frac{1}{n}(g^n(x) - x)$. Let $\| \cdot \|$ be one of the separating family of semi norms on V and $B = \mathrm{lub}_{c \in C} \| c \|$ (which is finite since C is compact). Since both x and $g^n(x) \in C$ by the triangle inequality $\| g(y) - y \| \leq \frac{2B}{n}$. Because this is true for all n and the left side of this inequality is independent of n we see $\| g(y) - y \| = 0$ and since this holds for all semi norms $g(y) = y$ for all $g \in G$.   □

Our next corollary follows from Theorem 2.3.6 in exactly the same way as Corollary 2.3.5 follows from Theorem 2.3.4.

Corollary 2.3.7. *Let G be an abelian topological group acting jointly continuously on a compact metric space X. Then there exists a G-invariant probability measure μ on X.*

We remark that although Corollary 2.3.5 holds for arbitrary compact groups and not merely abelian ones, in Corollary 2.3.7 it is essential that G be abelian (see [106], p. 40).

Definition 2.3.8. Let C be a compact topological space and $f : C \to C$ be a continuous self map, or a semigroup of such self maps. We define an *invariant probability measure* μ on C of a function f, or a semigroup of such functions by $\mu(f^{-1}(E)) = \mu(E)$, for every Borel set E of C. For a semigroup S of such functions, this holds for all $f \in S$.

Of course, when S is a group the condition, $\mu(f^{-1}(E)) = \mu(E)$, just means invariance in the sense defined earlier.

Example 2.3.9. Let $\phi(x) = x^2$. ϕ is a homeomorphism of $X = [0,1]$ onto itself. For any $0 < a < 1$, taking $E = [0, a]$ we have $\phi([0, a]) \subseteq [0, a^2]$ and since $a^n \to 0$, any probability measure μ on X invariant under ϕ has support contained in $\{0, 1\}$. Now identify the end points making $[0, 1]$ into a circle. The invariant probability measure here has a *one point* support. Apply a rotation ψ to the circle and consider the (nonabelian) group G generated by these two homeomorphisms. This G has no invariant probability measure since the support of such a measure would have to be empty, but the total mass is 1.

Finally for completeness we state the fixed point theorems of Ryll-Nardzewski and Furstenberg-Namioka (see [82]).

Theorem 2.3.10. *Let V be a locally convex linear topological vector space, C a (nonempty) weakly compact convex subset of V and S a semigroup of weakly continuous affine maps leaving C invariant and acting distally on C. Then C has a common S fixed point.*

Actually Hahn's theorem also holds for semigroups. The main distinction between Ryll-Nardzewski's theorem and Hahn's is that the former works for *weakly* continuous affine maps while the latter requires continuous affine maps.

We now state the fixed point theorem of Furstenberg-Namioka.

Theorem 2.3.11. *Let V be a locally convex linear topological vector space, C a (nonempty) compact convex subset of V and S a semigroup of continuous affine maps leaving C invariant and acting distally on C. Then C supports an S-invariant probability measure μ.*

When C is metric this was proved in 1963 by Furstenberg in [45] and in the general case in 1972 by Namioka [82].

2.4   Amenable Groups

We first define an invariant mean as follows:

Definition 2.4.1. Let G be a locally compact topological group and $BC(G)$ be the bounded continuous functions on G. A *left invariant mean m* is an $\mathbb{R}$-linear map $m : BC(G) \to \mathbb{R}$ satisfying the following conditions:

1. $m(f) \geq 0$ if $f \geq 0$.

2. $m(1) = 1$.

3. $m(f_g) = m(f)$ for any $f \in BC(G)$ and left translate by $g \in G$.

To render our statements simpler we formalize things by making the following definition.

Definition 2.4.2. A locally compact topological group G is said to have the *fixed point property* if whenever G acts jointly continuously and affinely on a locally convex linear topological vector space V and leaves invariant a compact convex subset C of V, it has a fixed point in C.

This brings us to fixed point theorems for amenable groups and the equivalence of the following three conditions for a locally compact topological group G.

1. G has a left invariant mean.

2. G has the fixed point property.

3. Any compact G-space X, X has a G-invariant probability measure.

For its many interesting applications, particularly to weak containment in representation theory, the reader should consult Greenleaf [47]. The equivalence of conditions 1 and 2 is proved in [47] and also in [82]. That 1 implies 2 is Day's fixed point theorem. Here (Theorem 2.4.4) we prove the equivalence of conditions 2 and 3. The more difficult 3 implies 2 will constitute the fixed point theorem of this section.

2.4.1   Amenable Groups

Definition 2.4.3. We shall call G an *amenable group* if it satisfies any one of these equivalent conditions.

Theorem 2.4.4. *An amenable group G acting continuously and affinely on a locally convex linear topological vector space V and leaving a compact convex set C in V invariant has a G-fixed point in C. Conversely, if G has the fixed point property any compact G-space X has a G-invariant probability measure.*

Proof. Suppose G acts continuously and affinely on a locally convex linear topological vector space V and leaves a compact convex set C in V invariant. Then G acts on $M(C)$, the space of all finite signed measures on C leaving invariant $M(C)^+$, the positive measures on C. This action of G on $M(C)$ is defined as follows: If μ is such a measure and E is a Baire set in C, we define $(g_*(\mu))(E) = \mu(g \cdot E)$ which clearly gives a finite signed measure and a G action on $M(C)$. If μ is positive, so is $g_*(\mu)$ so G stabilizes the probability measures.

Suppose G is amenable and C is a compact G-space. By 3, C supports a G-invariant probability measure μ. Let $b(\mu)$ be its *barycenter*. That is, $b(\mu)$ is the positive measure on C satisfying (and defined weakly by),

$$\lambda(b(\mu)) = \int_C \lambda(c) d(\mu(c)), \quad \lambda \in V^*.$$

If a measure μ were supported on a finite set $\{c_1, \ldots, c_n\}$, then $\mu = \sum_{i=1}^{n} t_i \delta_{c_i}$, where $t_i \geq 0$, and δ_{c_i} is the point mass at c_i. Then its barycenter $b(\mu) = \sum_{i=1}^{n} t_i \delta_{c_i}$, where $0 \leq t_i \leq 1$ and $\sum_{i=1}^{n} t_i = 1$. If g is any affine transformation in G of C, then $g_* b(\mu) = \sum_{i=1}^{n} t_i g_* \delta_{c_i} = \sum_{i=1}^{n} t_i \delta_{gc_i}$, where $0 \leq t_i \leq 1$ and $\sum_{i=1}^{n} t_i = 1$. While $g_*(\mu) = \sum_{i=1}^{n} t_i (\delta_{gc_i})$ so $b(g_*(\mu)) = \sum_{i=1}^{n} t_i (\delta_{gc_i})$, where also $0 \leq t_i \leq 1$ and $\sum_{i=1}^{n} t_i = 1$. Thus, for $g \in G$ and all positive measures μ with finite support

$$g_* b(\mu) = b(g_*(\mu)).$$

Since these measures are *dense* in $M(C)^+$ (see [33]) with its natural metric, by continuity this equation extends to all of $M(C)^+$. Now our particular μ is G-invariant. Hence $b(\mu)$ is G-fixed and its support is therefore also G-fixed. It is also nonempty. This is because by Choquet's theorem (see [25]) the support of $b(\mu)$ is the set of extreme points of C and by the Krein-Milman theorem (see [33], p. 440) the compact convex subset C of V has extreme points.

For the converse (which is very much like our proof of the existence of Haar measure on a compact group) suppose G has the fixed point property and X is a compact G-space. Then one checks easily that the action of G on $(M(X), w^*)$, the space of all signed measures on X, is affine and $M(X)^+$ is w^* compact by the theorem of Alaoglu (see [33]). It is certainly also convex. By the fixed point property there is a G-fixed point in $M(X)^+$, that is, a G-invariant probability measure on X.   □

So for example, if G is compact any continuous affine action on such a V has the fixed point property by the Kakutani fixed point theorem (Theorem 2.3.4). Similarly if G is abelian, any continuous affine action on such a V has the fixed point property by the Markov-Kakutani fixed point theorem (Theorem 2.3.6). Thus compact and abelian groups are amenable. Looked at another way, a compact group has a left invariant mean, namely Haar measure.

Curiously for a *finite dimensional* spaces, V, the fixed point property actually holds for *all semisimple groups* and in particular, for a noncompact simple group (see Theorem 1.4.14). But as we shall see in Proposition 2.4.6 these are definitely nonamenable. Thus the *infinite* dimensional character of V in the second condition is critical.

We mention some simple, but quite important facts. These are Theorems 2.3.1, 2.3.2 and 2.3.3 proved in [47].

Lemma 2.4.5. *Let G be a locally compact topological group and H a closed subgroup. Then,*

1. *If G is amenable so is H.*

2. *If N is a closed normal subgroup and G is amenable so is G/N.*

3. *If N a closed normal subgroup and both N and G/N are amenable so is G.*

Applying Lemma 2.4.5 in the abelian case yields the conclusion that any *solvable* locally compact group is amenable.

2.4.2 Structure of Connected Amenable Lie Groups

By Lemma 2.4.5 if a locally compact topological group G has a closed normal solvable subgroup R with G/R is compact, then G is amenable. In particular by the second isomorphism theorem, a locally compact group of the form $K \cdot R$, where K is a compact subgroup of G and R is a closed normal solvable subgroup of G is amenable.

We remark that if R is closed, normal and *simply connected*, and G/R is compact, then G is isomorphic to a semidirect product of K acting on R (see [1], Corollary 2.5.7) and so in particular $G = KR$.

An example of how amenable groups arise in nature is: Let G be a *real* semisimple Lie group of noncompact type, $G = KAN$ be its Iwasawa decomposition and M be the centralizer of A in K. Then MAN is a subgroup of G containing the connected solvable group AN

as a normal subgroup of finite index. Since multiplication gives a diffeomorphism of G with the product manifold $K \times A \times N$, MAN is closed and therefore locally compact. Thus MAN is amenable and so has the fixed point property. It is called a *Borel subgroup* of G.

In the case of a connected Lie group we have a complete characterization of amenability.

Proposition 2.4.6. *If G is a connected Lie group and* $\mathrm{Rad}(G)$ *is its radical, then G is amenable if and only if $G/\mathrm{Rad}(G)$ is compact.*

In particular by the Levi decomposition theorem, if G is a connected amenable Lie group, $G = \mathrm{Rad}(G)K$ for some compact (semisimple) K.

Proof. By the above, it is sufficient to prove if G is amenable, then $G/\mathrm{Rad}(G)$ is compact. That is, we must show that if a semisimple Lie group G is amenable, then it is compact. But such a group G is an almost direct product of a finite number of simple groups each of which is also amenable. Suppose at least one of these simple groups, say H, is noncompact. By Furstenberg's lemma $\mathrm{Ad}(H)$ (see [106], p. 39, or [1], p. 380) which acts on its Lie algebra $\mathfrak{h}$ and therefore also on the associated projective space, $P(\mathfrak{h})$, will not leave any measure on $P(\mathfrak{h})$ invariant. Hence by Theorem 2.4.4, $\mathrm{Ad}(H)$ is not amenable and therefore neither is H, a contradiction. Thus $G/\mathrm{Rad}(G)$ is compact. $\qquad\square$

We leave to the reader the verification of the fact that a connected Lie group has a unique largest normal amenable subgroup. Namely, $\mathrm{Rad}(G)K$, where K is the compact part of a Levi factor. Also, that if G is a Lie group and H is a closed subgroup of G with G/H having a finite G-invariant measure, then G is amenable if and only if H is amenable. In particular, *lattices* in noncompact semisimple groups are not amenable.

In this connection we should mention a result of von Neumann: A discrete group Γ is amenable if and only if it contains no nonabelian free group on two generators. Since lattices in Lie groups are always finitely generated and the Lie groups themselves are frequently linear, in view of the remarks just above, the well known Tits alternative (see [99]) comes into play. Namely, a finitely generated linear group Γ over a field of characteristic zero either contains a solvable subgroup of finite index (we say G is *virtually solvable*), or it contains a nonabelian free subgroup on two generators. If the group contains a solvable subgroup of finite index it must also contain a normal solvable subgroup of finite index and so is amenable. So a finitely generated linear group it is amenable if and only if it has a normal solvable subgroup of finite index and a lattice in a noncompact semisimple group always contains a free group.

Chapter 3

The Lefschetz Fixed Point Theorem

Our preliminary version of this theorem, which generalizes the Brouwer theorem, gives a cohomological (respectively homological) condition to insure that a continuous map f of a compact polyhedron X has a fixed point. To formulate this result we must first define the Lefschetz number of f. Initially we will work over $\mathbb{Q}$. Later we will find it convenient to move over to $\mathbb{R}$.

Denote by $f^{\star,k} : H^k(X) \to H^k(X)$ the induced map on the various cohomology groups, $H^k(X, \mathbb{Q})$. The Lefschetz number $L(f)$ is defined by:

$$L(f) = \sum_{k=0}^{n} (-1)^k \operatorname{Tr}\left(f^{\star,k} : H^k(X, \mathbb{Q}) \to H^k(X, \mathbb{Q})\right).$$

We remark that many authors define the Lefschetz number using the *homology* groups instead. They take

$$L(f) = \sum_{k=0}^{n} (-1)^k \operatorname{Tr}\left(f_{\star,k} : H_k(X, \mathbb{Q}) \to H_k(X, \mathbb{Q})\right),$$

where $H_k(X, \mathbb{Q})$ is the kth homology group of X with coefficients in $\mathbb{Q}$. These two definitions are equivalent by the universal coefficient theorem because the split exact sequence

$$(0) \to \mathrm{Ext}(H_k(X), \mathbb{Q}) \to H^k(X, \mathbb{Q}) \to \mathrm{Hom}(H_n(X), \mathbb{Q}) \to (0)$$

has the second term zero ($\mathrm{Ext}(A, B) = 0$ whenever B is a divisible group). In our situation, since we are working over a field ($\mathbb{Q}$, or $\mathbb{R}$), this fits well. When $f_\star$ acts on $H_k(X, \mathbb{Q})$ by a matrix A, $f^\star$ acts on $H^k(X, \mathbb{Q})$ by the transpose matrix A^t and since $\mathrm{Tr}(A) = \mathrm{Tr}(A^t)$ these definitions coincide.

In a later section we shall assume X is a compact smooth manifold and f a smooth self map of X with only *nondegenerate* fixed points. We will then show the number $\Lambda(f)$ of fixed points of f (counted with multiplicity) equals $L(f)$. As a compact manifold X can be triangulated. This follows from Whitney's imbedding theorem (see [79], p. 501) since such an X can be smoothly imbedded in $\mathbb{R}^n$ for some sufficiently high n. For the same reason its cohomology groups are finitely generated and vanish in high dimension.

3.1   The Lefschetz Theorem for Compact Polyhedra

Now $f^\star$ depends only on the homotopy class of f and hence $L(f)$ is a homotopy invariant. When $f = I_X$, the identity map of X, $L(f) = \chi(X)$, the Euler characteristic of X. This is because $\mathrm{Tr}(f^{*,k}) = \mathrm{Tr}(I_{H^k(X)}) = \dim(H^k) = k$ and so

$$L(f) = \sum_{k=0}^{n} (-1)^k \dim(H^k(X)) = \chi(X).$$

Actually, these traces are always integers.

Proposition 3.1.1. *Let A be a finitely generated abelian group and $g : A \to A$ be a group homomorphism Then $\mathrm{Tr}(g) \in \mathbb{Z}$.*

Proof. Let $g \otimes I : A \otimes \mathbb{Q} \to A \otimes \mathbb{Q}$ be the induced $\mathbb{Q}$-linear transformation. If T is the torsion subgroup of A, then g induces $\widetilde{g} : A/T \to A/T$

where A/T is a free abelian group of finite rank. Since $T \otimes \mathbb{Q} = 0$, the epimorphism $A \otimes \mathbb{Q} \to (A/T) \otimes \mathbb{Q}$ is an isomorphism, so we may identify $g \otimes I$ with $\tilde{g} \otimes I$. Choose a basis for A/T over $\mathbb{Z}$ which is also a basis for $(A/T) \otimes \mathbb{Q}$ over $\mathbb{Q}$. Hence, the matrix of $\tilde{g} \otimes I$ will be the same as the (integer) matrix of $\tilde{g}$ and so its trace will be an integer.   □

In its most primitive form the Lefschetz fixed point theorem states:

Theorem 3.1.2. *Let X be a compact polyhedron and $f : X \to X$ be a continuous function. If $L(f) \neq 0$, then f has a fixed point.*

One sees easily that the converse, that is if $L(f) = 0$, then f has no fixed points, is false. For example take f to be the identity map on a space X whose Euler characteristic is 0 such as the two torus. However, we remark that in [21], where $L(f)$ is defined with respect to any field F, the author proves a converse of the Lefschetz theorem under certain conditions on X. Namely, if $L(f, F) = 0$ for *all fields* F, then there is a map g homotopic to f with no fixed points.

Of course, Theorem 3.1.2 immediately recaptures Brouwer's fixed point theorem.

Corollary 3.1.3. (Brouwer's fixed point theorem). *Any continuous map of the closed unit ball B^n in $\mathbb{R}^{n+1}$ has a fixed point.*

Indeed, since $X = B^n$ is contractible, $L(f) = 1$ because the only nonzero cohomology group is $H^0(B^n) = \mathbb{Q}$.

For the proof of Theorem 3.1.2 we need the following simple lemma.

Lemma 3.1.4. *Given a commutative diagram with exact rows where*

$$
\begin{array}{ccccccccc}
0 & \longrightarrow & A & \xrightarrow{i} & B & \xrightarrow{\pi} & C & \longrightarrow & 0 \\
  & & \downarrow{f_1} & & \downarrow{f} & & \downarrow{f_2} & & \\
0 & \longrightarrow & A & \xrightarrow[i]{} & B & \xrightarrow[\pi]{} & C & \longrightarrow & 0,
\end{array}
$$

A, B, and C are finitely generated abelian groups then,

$$\mathrm{Tr}(f_1) + \mathrm{Tr}(f_2) = \mathrm{Tr}(f).$$

Proof. Tensoring by $\mathbb{Q}$ we again obtain an exact sequence of $\mathbb{Q}$-vector spaces.

$$0 \longrightarrow A \otimes \mathbb{Q} \xrightarrow{i \otimes id} B \otimes \mathbb{Q} \xrightarrow{\pi \otimes id} C \otimes \mathbb{Q} \longrightarrow 0$$

So, we have a direct sum decomposition, $B \otimes \mathbb{Q} = (A \otimes \mathbb{Q}) \oplus (C \otimes \mathbb{Q})$. Choosing an appropriate basis we get,

$$f = \left(\begin{array}{c|c} f_1 & \star \\ \hline 0 & f_2 \end{array} \right) .$$

Taking traces yields the result. $\square$

Our strategy in proving the Lefschetz theorem is to consider the same alternating sum of traces, but on the level of cochains, $C^k(X)$, and then show this is just $L(f)$. This brings us to Hopf's trace formula, Proposition 3.1.5 (where Z^k denotes the cocycles in C^k and B^k the coboundaries).

Proposition 3.1.5.

$$L(f) = \sum_{k=0}^{n} \text{Tr} \left(f^{\star,k} : C^k(X) \to C^k(X) \right).$$

Proof. We have the following diagrams for short exact sequences of finite dimensional $\mathbb{Q}$-vector spaces:

$$\begin{array}{ccccccccc}
0 & \longrightarrow & Z^k & \longrightarrow & C^k & \longrightarrow & B^{k+1} & \longrightarrow & 0 \\
 & & \downarrow f_1 & & \downarrow f & & \downarrow f_2 & & \\
0 & \longrightarrow & Z^k & \longrightarrow & C^k & \longrightarrow & B^{k+1} & \longrightarrow & 0
\end{array} \qquad (3.1)$$

and

$$0 \longrightarrow B^k \longrightarrow Z^k \longrightarrow H^k(X) \longrightarrow 0$$

$$f_1^{\star,k} \downarrow \qquad f^{\star,k} \downarrow \qquad f_2^{\star,k} \downarrow \qquad (3.2)$$

$$0 \longrightarrow B^k \longrightarrow Z^k \longrightarrow H^k(X) \longrightarrow 0$$

where, in each diagram, the first vertical arrow is the restriction of f (resp. of $f^{\star,k}$), and the last, the induced quotient map. Applying Lemma 3.1.4 to each diagram tells us for every k,

$$\mathrm{Tr}(f^{\star,k}|_{Z^k}) + \mathrm{Tr}(f^{\star,k}|_{B^{k+1}}) = \mathrm{Tr}(f^{\star,k}), \qquad (3.3)$$

and

$$\mathrm{Tr}(f^{\star,k}|_{B^k}) + \mathrm{Tr}(f^{\star,k}|_{H^k(X)}) = \mathrm{Tr}(f^{\star,k}). \qquad (3.4)$$

Now, set

$$B_t = \sum_k t^k \,\mathrm{Tr}(f^{\star,k}|_{B^k}), \quad C_t = \sum_k t^k \,\mathrm{Tr}(f^{\star,k}|_{C^k}),$$

and

$$H_t = \sum_k t^k \,\mathrm{Tr}(f^{\star,k}|_{H^k(X)}), \quad Z_t = \sum_k t^k \,\mathrm{Tr}(f^{\star,k}|_{Z^k}),$$

where $t \in \mathbb{R}$. Using (3.3) we get

$$C_t = Z_t + \frac{1}{t} B_t$$

and by (3.4)

$$H_t = Z_t - B_t.$$

Subtracting,

$$C_t - H_t = \frac{1+t}{t} B_t.$$

Setting $t = -1$, we see $C_{-1} = H_{-1}$. $\square$

Proof of Lefschetz's theorem.

Proof. Let X be compact and suppose f has no fixed points. Then f must move each point x a minimum distance $\epsilon > 0$ (independent of x). By considering simplicial cochains $C^k(X)$, we can subdivide X finely enough into simplices so that each of the diameters is less than $\frac{\epsilon}{4}$. By the simplicial approximation theorem f can be approximated by a *simplicial* map g which is everywhere within $\frac{\epsilon}{2}$ of f. Because ϵ is small and g is homotopic to f, without loss of generality we can assume g is f. Since any of these simplices, σ, has diameter less than $\frac{\epsilon}{4}$ and f moves points a distance at least ϵ, $\sigma \cap f^\star(\sigma) = \emptyset$. But as we saw above, the matrix of $f^{\star,k}$ has only integer entries. This forces its diagonal to consist only of 0's. Hence for any k, $\mathrm{Tr}(f^{\star,k}) = 0$. By Hopf's Trace Theorem 3.1.5, $L(f) = 0$, a contradiction.   □

3.1.1   Projective Spaces

Here we use the Lefschetz's theorem to calculate fixed points of continuous quasi-linear mappings of projective space $FP^{n-1} \subseteq F^n$, where $F = \mathbb{R}$, or $\mathbb{C}$.

We first deal with the real case. Consider an invertible linear transformation $f : \mathbb{R}^n \to \mathbb{R}^n$. Since f is linear it takes lines through 0 to lines through 0 and hence induces the *quasi-linear map* $\widetilde{f} : \mathbb{R}P^{n-1} \to \mathbb{R}P^{n-1}$. As we know,

$$H^k(\mathbb{R}P^n,\ \mathbb{Q}) = \begin{cases} \mathbb{Q} & k = 0, \\ \mathbb{Q} & k = n, \quad n \text{ odd}, \\ 0 & k = n, \quad n \text{ even}, \\ 0 & k \neq 0, \quad n. \end{cases}$$

Therefore, $\chi(\mathbb{R}P^n) = \begin{cases} 1, & n \text{ even} \\ 0, & n \text{ odd} \end{cases}$, which means when n is even $\widetilde{f}$ has a fixed point.

Now, in general, fixed points of $\widetilde{f}$ are eigenvectors of f. For n odd since the characteristic polynomial of f has odd degree and hence has a real root. Thus an eigenvector exists in this case which is in agreement with the observation above that every map $\mathbb{R}P^{2k} \to \mathbb{R}P^{2k}$ has a fixed point.

On the other hand, the rotation of $\mathbb{R}^{2k}$ defined by $f(x_1, ..., x_{2k}) = (x_2, -x_1, x_4, -x_3, ..., x_{2k}, -x_{2k-1})$ has no eigenvectors and so its projectivization $\tilde{f} : \mathbb{R}P^{n-1} \to \mathbb{R}P^{n-1}$ has no fixed points.

As a corollary of the Lefschetz theorem in the case of complex projective space we have,

Corollary 3.1.6. *Every continuous map $f : \mathbb{C}P^{2n} \to \mathbb{C}P^{2n}$ has a fixed point.*

Proof. We know that the cohomology ring of $\mathbb{C}P^j$ is given by $\mathbb{Z}[x]/(x^j + 1)$, where x is a generator of $H^2(\mathbb{C}P^j)$. So, if

$$f^{\star,2} : H^2(\mathbb{C}P^j) \to H^2(\mathbb{C}P^j)$$

is the multiplication by m, then (by the cup product),

$$f^{\star,k} : H^{2k}(\mathbb{C}P^j) \to H^{2k}(\mathbb{C}P^j)$$

is multiplication by m^k (here $f^{\star,k}(x^k) = (f^{\star,2}(x))^k = (mx)^k = m^k x^k$, since $f^{\star,k}$ is a ring homomorphism). Therefore, for $j = 2n$,

$$L(f) = \sum_{k=0}^{2n}(-1)^k \operatorname{Tr}(f^{\star,k}) = 1 + m + m^2 + \cdots + m^{2n}$$

$$= \begin{cases} \frac{1-m^{2n+1}}{1-m} & \text{if } m \neq 1 \\ 2n+1 & \text{if } m = 1 \end{cases}$$

showing $L(f) \neq 0$. $\square$

This calculation yields,

Corollary 3.1.7. *Every continuous map $f : \mathbb{C}P^n \to \mathbb{C}P^n$ for which*

$$f^{\star,2} : H^2(\mathbb{C}P^j) \to H^2(\mathbb{C}P^j)$$

is multiplication by $m \neq -1$ has a fixed point.

3.2   The Lefschetz Theorem for a Compact Manifold

3.2.1   Preliminaries from Differential Topology

For the reader's convenience we review some notions concerning manifolds. In particular, we will define the tangent and cotangent bundles, the normal bundle, tubular neighborhoods, de Rham cohomology, orientation and finally transversality. These concepts will be of use in Chapter 4 as well. For the most part we shall be interested in *compact manifolds* and for this reason the existence of these various objects is no problem at all. For the reader unfamiliar with the material of this section, rather than plow through it, we recommend returning to this section as needed.

Let M be an n-dimensional smooth manifold, $T_p(M)$ the *tangent space* at a point $p \in M$, and $\frac{\partial}{\partial x_1}, \ldots, \frac{\partial}{\partial x_n}$ its basis in a local chart $(U, x_1, \ldots, x_n)$ at p.

Now suppose X is a k-dimensional submanifold of M where $k < n$. Let $i : X \hookrightarrow M$ be the inclusion map. At each $x \in X$, the tangent space to X is viewed as a subspace of the tangent space to M via the linear inclusion $di_x : T_x(X) \hookrightarrow T_x(M)$, where we identify x and $i(x)$. The quotient $N_x(X) := T_x(M)/T_x(X)$ is an $(n-k)$-dimensional vector space, known as the *normal space* to X at x. The *normal bundle* of X is

$$N(X) = \{(x, v) : x \in X, v \in N_x(X)\}.$$

Under the natural projection $N(X)$ can be given the structure of a vector bundle over X of rank $n - k$. Hence, as a manifold, $N(X)$ is n-dimensional. The zero section of $N(X)$,

$$i_0 : X \hookrightarrow N(X), \quad x \mapsto (x, 0),$$

imbeds X as a closed submanifold of $N(X)$. A neighborhood U of the zero section X in $N(X)$ is called *convex* if the intersection, $U \cap N_x(X)$, with each fiber is convex.

Just as a smooth curve (with finite parameter interval) $c(t)$ in $\mathbb{R}^3$ is contained in a neighborhood shaped like a tube, a submanifold M of N

also always has a tubular shaped neighborhood about it. Now if M is compact, it can be imbedded in $N = \mathbb{R}^n$ for some n. Let $d = |\cdot|$ be the usual distance in $\mathbb{R}^n$. We define a tubular neighborhood of M as follows:

Definition 3.2.1. A *tubular neighborhood* is

$$V_\epsilon = \{x \in \mathbb{R}^n \mid d(x; M) < \epsilon\}.$$

For $x \in \mathbb{R}^n$ let $c(x) \in M$ to be the closest point to x in M (for ϵ sufficiently small, any $x \in V_\epsilon$ has a unique nearest point $c(x) \in M$, i.e. a unique point $c(x) \in M$ minimizing the distance from x to $c(x)$).

When M is compact, tubular neighborhoods satisfy the following:

Theorem 3.2.2. (Tubular neighborhood theorem). *Let M be a smooth compact manifold. For ϵ small enough,*

$$\pi : (V_\epsilon,\ \partial V_\epsilon) \to M$$

defined by $x \mapsto c(x)$ is a smooth and surjective.

We omit the proof of this result, which can be found in [27], pp. 43, 61.

We denote by $T_p^\star(M) = \operatorname{Hom}(T_p(M), \mathbb{R})$ the *cotangent space* and $dx_1, ..., dx_n$ the corresponding basis of $T_p^\star$. For each k, we consider the space $\bigwedge^k T_p^\star(M)$ of dimension $\binom{n}{k}$. A basis for this space is

$$\{dx_{i_1} \wedge \cdots \wedge dx_{i_k} \mid i_1 < i_2 < \cdots < i_k\}.$$

Definition 3.2.3. For $k \geq 1$, a *k-form* on M is an assignment of an element $\omega(p) \in \bigwedge^k T_p^\star(M)$ to each $p \in M$ depending smoothly on p. A 0-form is just a smooth function $f : M \to \mathbb{R}$.

If ω is a k-form, the exterior derivative $d\omega$ is the $(k+1)$-form locally defined on a chart $(U, x_1, ..., x_n)$ by

$$d(f\, dx_{i_1} \wedge \cdots \wedge dx_{i_k}) = \sum_{i=1}^{n} \frac{\partial f}{\partial x_i} dx_i \wedge dx_{i_1} \wedge \cdots \wedge dx_{i_k}.$$

When $k = 0$, $df(v) = vf$. Also,

$$d(\alpha \wedge \beta) = d\alpha \wedge \beta + (-1)^{\deg \alpha}\alpha \wedge d\beta.$$

Since locally, $\frac{\partial^2}{\partial x_i \partial x_j} = \frac{\partial^2}{\partial x_j \partial x_i}$ and $dx_i \wedge dx_j = -dx_j \wedge dx_i$, one sees easily that $d^2 = 0$ so we have a complex called the *de Rham complex*.

Now let

$$\Omega^k(M) = \{k - \text{forms on } M\}$$

(in other words $\Omega^k(M)$ is the space $\Gamma(\Lambda^k T^\star(M))$ of smooth sections of the kth exterior power of the cotangent bundle).

$$0 \longrightarrow \Omega^0 \xrightarrow{\ d\ } \Omega^1 \xrightarrow{\ d\ } \cdots \xrightarrow{\ d\ } \Omega^n \longrightarrow 0.$$

The operator $d : \Omega^0(M) \to \Omega^1(M)$ is simply $df(X) = X(f)$, where f is a smooth function and X is a tangent vector acting on functions as the directional derivative.

We are now in a position to define de Rham cohomology.

Definition 3.2.4. The k-de Rham cohomology group as

$$H^k(M) = \text{Ker}(d : \Omega^k \to \Omega^{k+1})/\text{Im}(d : \Omega^{k-1} \to \Omega^k).$$

For example, if $M = \mathbb{R}$, then $d : \Omega^0 \to \Omega^1$ and $f \mapsto f'dx$. Therefore,

$$H^0(\mathbb{R}) = \text{Ker}(d : \Omega^0 \to \Omega^1) = \mathbb{R},$$

and since $gdx = df$ with $f(x) = \int_0^x g(t)dt$,

$$H^1(\mathbb{R}) = \Omega^1/d(\Omega^0) = \Omega^1/\Omega^1 = (0).$$

Another notion we shall require is that of *orientation* of a manifold. We begin with a fixed *ordered* basis $\{e_1, \ldots, e_n\}$ of $\mathbb{R}^n = V$. Consider a linear transformation T of V and its matrix A with respect to this basis. Changing the order of $\{e_1, \ldots, e_n\}$ by a permutation, σ, changes $\det A$ by multiplying by $\text{sgn}(\sigma)$. But, of course, we will not see this unless $\det(T) \neq 0$. So now restrict to T (or A) in $\text{GL}(n, \mathbb{R})$. Since $\text{sgn}(\sigma) = \pm 1$ we partition $\text{GL}(n, \mathbb{R})$ respectively into those transformations with positive determinant $\text{GL}^+(n, \mathbb{R})$, (resp. negative) determinant $\text{GL}^-(n, \mathbb{R})$.

Definition 3.2.5. By an orientation of $\mathbb{R}^n$ we mean the choice of $GL^+(n, \mathbb{R})$ or $GL^-(n, \mathbb{R})$ as the preferred source for picking a basis for $\mathbb{R}^n$. The standard orientation for $\mathbb{R}^n$ is that determined by $GL^+(n, \mathbb{R})$. Given a manifold M and a point $p \in M$, this gives an orientation of $T_p(M)$. An orientation on M itself is an orientation of each $T_p(M)$, $p \in M$, which *varies continuously* with p.

Evidently, this definition only has significance on each connected component of M.

Given a map $f : M \to N$, a point $x \in M$ is called *regular* if the linear map df_x is nondegenerate. By the inverse function theorem f is a diffeomorphism in a neighborhood of x. A point is *singular*, if it is not regular. A point $y \in N$ is called a *regular value* if $f^{-1}(y)$ consists only of regular points or is empty, and a singular value otherwise.

Proposition 3.2.6. *If M is compact, $f : M \to N$ and $y \in N$ is a regular value, then $f^{-1}(y)$ is a finite set.*

Proof. The set $f^{-1}(y)$ is a closed since f is continuous and it therefore compact since M is. Moreover, for each $x \in f^{-1}(y)$ the map f is a diffeomorphism of a sufficiently small neighborhood of x to a neighborhood of y. Therefore the set $f^{-1}(y)$ is also discrete. As a discrete and compact set it is finite. $\square$

3.2.2   Transversality

Definition 3.2.7. Let M be a manifold. We say that two submanifolds A and B of M *intersect transversely*[1], and we write $A \pitchfork B$, if at every point $p \in A \cap B$, $T_p(M) = lin.sp.(T_p(A), T_p(B))$.

The notion of transversality generalizes that of a regular value of a function. Indeed, if $f : M \to N$ is a smooth map of manifolds and M happens to be a point, p, then f is transversal to p if and only if p is a *regular value* of f.

---

[1] The notion of a "transverse intersection" seems to originate in R. Thom's thesis in the early fifties in [98].

More generally, if N_1, N_2, M are manifolds and $f_1 : N_1 \to M$, $f_2 : N_2 \to M$ are smooth maps, we say f_1 and f_2 are *transverse* if

$$f_{1\star}(T_{x_1}(N_1)) \oplus f_{2\star}(T_{x_2}(N_2)) = T_y(M),$$

for all $x_1 \in N_1$, $x_2 \in N_2$, with $f_1(x_1) = y = f_2(x_2)$.

Moreover, if $f : N \to M$ is a smooth map and $S \subseteq M$, an imbedded submanifold, we say f is *transversal* to S if for each $x \in f^{-1}(S)$, $f_\star(T_x(N))$ and $T_{f(x)}(S)$ together span $T_{f(x)}(M)$.

Proposition 3.2.8. *If A and B intersect transversely, $A \cap B$ is also a submanifold.*

Proof. The exponential map (for some Riemannian metric) furnishes an open neighborhood U of every point $p \in A \cap B$ such that

$$(U; U \cap A, U \cap B) \cong (\mathbb{R}^n, V_1, V_2)$$

for transverse linear subspaces V_1, V_2.   □

Equivalently, A and B intersect transversely if and only if

1. $A \cap B$ is a submanifold of M, and

2. for any $p \in A \cap B$, there is a short exact sequence

$$0 \to T_p(A \cap B) \to T_p(A) \oplus T_p(B) \to T_p(M) \to 0.$$

Now suppose $\dim(M) = n$, $\dim(A) = n - i$ and $\dim(B) = n - j$. Then $\dim(A \cap B) = n - (i + j)$ and we adopt the following convention for the orientation of $A \cap B$. Choose an oriented basis $u_1, ..., u_{n-(i+j)}, v_1, ..., v_i, w_1, ..., w_j$ for $T_p(M)$ so that $u_1, .., u_{n-(i+j)}, v_1, ..., v_i$ is an oriented basis for $T_p(A)$ and $u_1, ..., u_{n-(i+j)}, w_1, ..., w_j$ is an oriented basis for $T_p(B)$. We declare the basis $u_1, ..., u_{n-(i+j)}$ to be an oriented basis for $T_p(A \cap B)$.

In the case of a compact oriented manifold M and A and B two oriented submanifolds of dimensions k and l, respectively, with $k + l = n$, and which intersect transversely, this orientation on $A \cap B$ means that for a point $p \in A \cap B$ the canonical isomorphism

$$T_p(A) \oplus T_p(B) \simeq T_p(M)$$

is *orientation preserving*.

Now, let M be a *compact* manifold, and $f : M \to M$ be a smooth function.

Definition 3.2.9. A fixed point p of f is *nondegenerate* (or of *Lefschetz type*) if the differential $df_p : T_p(M) \to T_p(M)$ does not have 1 as an eigenvalue. In other words, $\det(\mathrm{I} - df_p) \neq 0$, where I is the identity map on $T_p(M)$.

In this case we define the *index* or the *Lefschetz sign* of the fixed point p by

$$\nu_p = \begin{cases} 1 & \text{if } \det(\mathrm{I} - df)_p > 0 \\ -1 & \text{if } \det(\mathrm{I} - df)_p < 0. \end{cases}$$

Using the Inverse Function Theorem one sees easily that nondegenerate fixed points are isolated and finite in number (for more details see [79], p. 218). Let $\mathrm{Fix}(f)$ denote the set of fixed points of f.

Definition 3.2.10. If all fixed points of f are nondegenerate, we define

$$\Lambda(f) = \Sigma_{p \in \mathrm{Fix}(f)} \nu_p \in \mathbb{Z}.$$

As usual, the *graph* of f, is $\Gamma_f = \{(p, f(p)) \mid p \in M\} \subseteq M \times M$, so Γ_f is a submanifold of $M \times M$ and the map

$$1_M \times f : M \to M \times M, \quad p \mapsto (p, f(p)),$$

is an embedding with $\mathrm{Im}(1_M \times f) = \Gamma_f$. This map gives an orientation on Γ_f.

Definition 3.2.11. In the special case where $f = I_M$, $\Gamma_{I_M} = \Delta \subset M \times M$, is called the *diagonal* of M.

Proposition 3.2.12. *Let M be a compact oriented manifold M and $f : M \to M$ a smooth map. The map f has only nondegenerate fixed points if and only if Γ_f intersects Δ transversely.*

Proof. We identify $\text{Fix}(f)$ and $\Gamma_f \cap \Delta$. Let $p \in \text{Fix}(f)$ and $e_1, \ldots, e_n$ be an oriented basis of $T_p(M)$. Then, $(e_1, e_1), \ldots, (e_n, e_n)$ is an oriented basis of $T_{(p,p)}(\Delta)$, and $(e_1, df_p(e_1)), \ldots, (e_n, df_p(e_n))$ is an oriented basis of $T_{(p,p)}(\Gamma_f)$. By definition

$$\Gamma_f \pitchfork \Delta \iff \text{these } 2n \text{ vectors are linearly independent.}$$

But since,

$$\det \left(\begin{array}{c|c} \text{I} & \text{I} \\ \hline df_p & \text{I} \end{array} \right) = \det \left(\begin{array}{c|c} \text{I} - df_p & 0 \\ \hline df_p & \text{I} \end{array} \right) = \det(\text{I} - df_p),$$

this occurs if and only if $\det(\text{I} - df_p) \neq 0$. Thus, $\Gamma_f \pitchfork \Delta$ if and only if p is a nondegenerate point. $\qquad\square$

This suggests that for a compact manifold M, we call the map $f : M \to M$ *transversal* if $\Gamma_f \pitchfork \Delta$ in $M \times M$.

Definition 3.2.13. An open cover $(U_i)_{i \in I}$ of M is called a *good cover* if all finite intersections

$$U_{i_1} \cap U_{i_2} \cap \ldots \cap U_{i_l}$$

are diffeomorphic to $\mathbb{R}^n$.

We have the following crucial proposition (see [16], p. 42).

Proposition 3.2.14. *Every compact manifold has a finite good cover.*

This proposition has very important implications. For example, for a compact manifold, as a consequence of Proposition 3.2.14 one gets the finite dimensionality of the de Rham cohomology, Poincaré duality, the Künneth formula and the Thom isomorphism.

For manifolds L and M the projection maps π_L and π_M from $L \times M$ to L and M respectively, induce morphisms of the cohomology rings

$$\pi_L^\star : H(L)^\star \to H^\star(L \times M) \text{ and } \pi_M^\star : H^\star(M) \to H^\star(L \times M).$$

As a result,

Theorem 3.2.15. (Künneth's Formula). *For compact manifolds L and M the map,*

$$(\pi_L^\star, \pi_M^\star) : H^\star(L) \otimes H^\star(M) \to H^\star(L \times M)$$

is a ring isomorphism.

For a proof see [16], pp. 47-50.

Actually, the de Rham cohomology is isomorphic to both Čech cohomology and singular cohomology (all with real coefficients). As a result we have the following theorem where the isomorphism is given by integration of differential forms over smooth singular chains.

Theorem 3.2.16. (de Rham Theorem). *For a compact manifold M there is a natural isomorphism*

$$H_{dR}^\star \cong H_{sing}^\star(M, \mathbb{R}).$$

We also have the *Poincaré duality theorem* as follows.

Theorem 3.2.17. *For an n-dimensional compact manifold M,*

$$H^k(M) = H^{n-k}(M)^\star,$$

where $0 \leq k \leq n$.

Let M be a compact oriented manifold of dimension n. The simplest place to find de Rham cohomology classes is in dimension n where there is a unique normalized volume form, $\int_M$. Thus if α is an n-form on M, we can form $\int_M \alpha \in \mathbb{R}$. If $\alpha = d\beta$, then by Stokes' theorem (see [79]),

$$\int_M \alpha = \int_M d\beta = 0.$$

Thus, $\int_M \alpha$ depends only on the cohomology class of α in $H^n(M)$. In particular if $\int_M \alpha \neq 0$, the cohomology class of α is nonzero.

Now let $p \in M$ and $(U; x_1, \ldots, x_n)$ be an oriented local coordinate chart around p, with $x_1(p) = \cdots = x_n(p) = 0$. Let $\varphi : \mathbb{R}^n \to \mathbb{R}$ be a smooth nonnegative function with small compact support near 0 and with $\int \varphi = 1$. Then the *bump* n-form

$$\alpha = \varphi(x_1, \ldots, x_n)dx_1 \wedge \cdots \wedge dx_n$$

is well-defined on M and vanishes outside a neighborhood of p. Also, $\int_M \alpha = 1$, so it defines a nonzero class in $H^n(M)$. It follows from the Poincaré duality theorem that for M connected, $H^n(M)$ is one-dimensional and generated by the cohomology class of any bump n-form.

We now turn to the concept of the *degree of a map* in the case of a compact, connected, oriented, n-dimensional manifold M without boundary.

Proposition 3.2.18. *There is an isomorphism $r : H^n(M) \to \mathbb{R}$ given by $r([\alpha]) = \int_M \alpha$.*

For a proof see [16], p. 40.

This enables us to define the degree of a smooth map $f : M \to N$, where M and N are two compact, oriented smooth n-dimensional manifolds. The map f induces[2] a linear map $f^\star : H^n(N) \to H^n(M)$, with the following commutative diagram,

$$
\begin{array}{ccc}
H^n(N) & \xrightarrow{\;f^\star\;} & H^n(M) \\
{\scriptstyle r}\downarrow & & \downarrow{\scriptstyle r} \\
\mathbb{R} & \xrightarrow[\lambda]{} & \mathbb{R},
\end{array}
$$

---

[2]If $f : M \to N$ is a smooth map and $\alpha \in \Omega^k(N)$ is a differential k-form on N, its *pullback* under f is the differential k-form $f^\star\alpha \in \Omega^k(M)$ defined by

$$(f^\star\alpha)_p(v_1, \ldots, v_k) := \alpha_{f(p)}(df_p(v_1), \ldots, df_p(v_k))$$

for $p \in M$ and $v_1, \ldots, v_k \in T_p(M)$.

where $\lambda \in \mathbb{R}$ is defined as $\lambda \int_N \alpha = \int_M f^\star(\alpha)$ for the generator α of $H^n(M)$. One can prove that λ depends only on the homotopy class of f.

Definition 3.2.19. The number λ is called the *degree* of f and is denoted by $\deg(f)$.

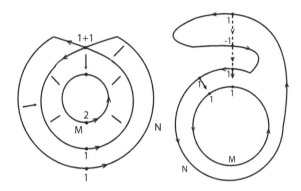

One can prove that if $q \in N$ is a regular value of f, then

1. $\deg(f) = n_+ - n_-$, where n_+ is the number of $p \in f^{-1}(q)$ for which df_p preserves orientation and n_- the number of $p \in f^{-1}(q)$ for which df_p reverses orientation. This means *that the degree of f is an integer.*

2. Homotopic maps have the same degree.

3. If $f : M \to N$ and $g : N \to P$, then

$$\deg(fg) = \deg(f) \cdot \deg(g).$$

   In particular $\deg I = 1$.

4. The degree of an orientation reversing diffeomorphism is -1. In particular an orientation reversing diffeomorphism cannot be homotopic to the identity map.

In the above construction from a point p, which is a 0-dimensional manifold, we got an n-dimensional cohomology class. We will now generalize this by starting with a higher-dimensional submanifold rather than a point.

Suppose M is a compact oriented n-manifold, and N is a compact oriented $(n - k)$-dimensional submanifold. By the Tubular Neighborhood Theorem 3.2.2, there is a neighborhood of N in M which is diffeomorphic to the total space of a k-dimensional oriented vector bundle over N, namely the normal bundle $(T(M)|_N)/T(N)$.

Now, pick a Thom form[3] for this vector bundle, and transfer it to the tubular neighborhood by the diffeomorphism and extend it (by setting it $= 0$) to a form on the whole of M. The result is a smooth differential form on M. Since the Thom form was compactly supported within the tubular neighborhood we obtain a closed k-form defining a cohomology class $[\alpha_N] \in H^k(M)$. When $k = n$ this was our earlier construction.

Proposition 3.2.20. *Let N_1 and N_2 be oriented submanifolds of M having dimensions $n - k$ and k respectively, and suppose that they meet transversely in a finite set of points. Then the integral*

$$\int \alpha_{N_1} \wedge \alpha_{N_2}$$

is the signed count of the number of points of intersection of N_1 and N_2.

For an oriented manifold M and an oriented submanifold, A, without boundary of codimension k, one defines the class $[A]$ in $H^\star(M)$ as the

---

[3]Let $E \to M$ be a vector bundle over M of rank n, and D be the unit disk bundle over M consisting of vectors v with $\| v \| \leq 1$. Define S as the unit sphere bundle over M of all vectors v with $\| v \| = 1$. By excision we get the canonical isomorphism $H^\star(E, E \setminus M, \mathbb{Z}) \cong H^\star(D, D \setminus M, \mathbb{Z})$. Now, by the long exact sequence of the triple $(D, D \setminus M, S)$, we have $H^\star(D, D \setminus M, \mathbb{Z}) \cong H^\star(D, S, \mathbb{Z})$. Therefore, $H^\star(E, E \setminus M, \mathbb{Z}) \cong H^\star(D, S, \mathbb{Z})$. The Thom class is the unique element of $H^\star(D, S, \mathbb{Z})$, which for each $p \in M$ restricts to the generator of $H^n(D_p, S_p, \mathbb{Z})$ determined by the orientation. Intuitively, the Thom class, η, evaluated on an n-chain, α, counts the intersections of α with the zero sections (the points of M).

class of an element $\eta_A \in \Omega^k(M)$ with $d\eta_A = 0$ such that for every $\omega \in \Omega_c(M)$ with $d\omega = 0$,

$$\int_A \omega = \int \omega \wedge \eta_A.$$

Now, let M be an oriented manifold and A and B two oriented submanifolds which intersect transversely. Let $i_B : B \to M$ be the inclusion map and $[A \cap B; B]$ be the class of $[A \cap B]$ in B. Then

Proposition 3.2.21. $[A \cap B; B] = i_B^*([A])$ in $H(B)$ and $[A \cap B] = [A] \cdot [B]$ in $H(M)$.

Proof. We first show that the second claim follows from the first. Let η_A and η_B in $\Omega(M)$ represent $[A]$ and $[B]$, respectively. According to the first claim, $i_B^* \eta_A$ represents $[A \cap B, B]$. Let ω be in $\Omega_c(M)$. Then

$$\int_M \omega \wedge \eta_A \wedge \eta_B = \int_B \omega \wedge \eta_A = \int_{A \cap B} \omega.$$

It remains to prove the first claim. To do this we will construct suitably related tubular neighborhoods of A in M and $A \cap B$ in B. First, choose a metric on $T(M)$. For x in $A \cap B$ we have

$$N_x(A \cap B) = N_x(A) \oplus N_x(B),$$

within $T_x(M)$, where now $N_x(A) = (T_x(A))^{\perp}$, $N_x(B) = (T_x(B))^{\perp}$, and $N_x(A \cap B) = (T_x(A \cap B))^{\perp}$. Let $N(A \cap B, B)$ be the normal bundle of $A \cap B$ in $T(B)|A \cap B$. Then, $N_x(A) \oplus N_x(B) = N_x(A \cap B) = N_x(B) \oplus N_x(A \cap B, B)$, which gives us an isomorphism $f : N(A)|_{A \cap B} \to N(A \cap B, B)$.

Let U be a sufficiently small tubular neighborhood of the zero section in $N(A)$ such that the exponential map $\phi : U \to M$ is defined on it, as well as the exponential map $\psi : f(U \cap N(A)|A \cap B) \to B$. Then ϕ is a tubular neighborhood of A in M, and ψ is one for $A \cap B$ in B. If U is sufficiently small, for each v in $U \cap N(A)|A \cap B$, there is a unique shortest geodesic from $\phi(v)$ to $\psi(fv)$, giving us a homotopy from ϕ to $\psi \circ f$. Now, choose an η in $\Omega(U)$ of degree equal to codim(A) which is closed, with support proper over A and satisfying $\int_{N_a(A)} \eta = 1$, for each a in A. The homotopy from ϕ to $\psi \circ f$ and the closedness of η show η has integral 1 over the fibers of the tubular neighborhood ψ.   □

Corollary 3.2.22. *If in addition M is a compact manifold and* $\dim A +$ $\dim B = \dim M$, *then* $A \cap B$ *is finite and*

$$\sum_{p \in A \cap B} \nu_p = \int_M [A] \cdot [B].$$

Proof. The dimension of $A \cap B$ is zero. As M is compact, so is $A \cap B$. Therefore $A \cap B$ is a finite set say $p_1, ..., p_n$. Note that $[A \cap B]$ is represented by the sum $\omega_1 + \cdots + \omega_n$, with each ω_i a volume form concentrated around p_i, of weight 1 if the orientations match, and weight -1 if they do not.   □

Now, let $f : M \to M$ be a smooth map. We want to count the number of fixed points p of f (with multiplicities). As we saw, these fixed points are precisely the points in $\Gamma_f \cap \Delta$, where Δ is the diagonal in $M \times M$. This number is given by Theorem 3.3.2.

3.3   Proof of the Lefschetz Theorem

Definition 3.3.1. We call a map f a *Lefschetz map* if all of its fixed points are of Lefschetz type.

Theorem 3.3.2. (Lefschetz Theorem). *Let M be a compact, orientable, manifold and $f : M \to M$ be a smooth map whose graph Γ_f is transversal to the diagonal Δ. Then,*

$$\Lambda(f) = L(f).$$

The proof of this fact will involve several steps. First, we will compute the cohomology classes $[\Gamma_f]$ and $[\Delta]$ in $H^\star(M)$. Then, we will compute $[\Gamma_f] \cdot [\Delta]$ as the right hand side of the above formula, and $[\Gamma_f \cap \Delta]$ as the left hand side. We will also choose an orientation of M, but the proof will be independent of this.

Proposition 3.3.3. *Let M be a compact oriented manifold and $f : M \to M$ be a smooth map. Let $[\omega_i]$ be a basis of $H^\star(M)$, with each ω_i homogeneous, say of degree $\deg \omega_i$. If $[\tau_i]$ is the Poincaré dual basis then*

$$\int_M \omega_i \wedge \tau_j = \delta_{ij}.$$

Let π_1, $\pi_2 : M \times M \to M$ be the two projections. Then, by Künneth's formula, $\pi_1^\star[\omega_i] \cdot \pi_2^\star[\tau_j]$, $i, j \in I$, form a basis of $H^\star(M \times M)$. Let the real numbers $f_{i,j}$, $(i,j) \in I^2$, be defined by

$$f^\star[\omega_i] = \sum_j f_{j,i}[\omega_j].$$

Then one has:

$$[\Gamma_f] = \sum_{i,j}(-1)^{\deg \omega_i} f_{i,j} \pi_1^\star[\omega_i] \cdot \pi_2^\star[\tau_j] = \sum_j (-1)^{\deg \omega_j} \pi_1^\star f^\star[\omega_j] \cdot \pi_2^\star[\tau_j].$$

In particular, for $f = I_M$ we have:

$$[\Delta] = \sum_i (-1)^{\deg \omega_i} \pi_1^\star[\omega_i] \cdot \pi_2^\star[\tau_i] = (-1)^{\dim M} \sum_i (-1)^{\deg \omega_i} \pi_2^\star[\omega_i] \cdot \pi_1^\star[\tau_i].$$

Proof. First we note that for a finite dimensional vector space V, with basis v_i and dual basis $v_i^\star$ and an endomorphism T of V

$$T = \sum_{i,j}(v_j^\star(T(v_i))) \cdot v_i^\star \otimes v_j.$$

Now let $\iota : M \to M \times M$ be the map taking $p \mapsto (p, f(p))$. There are unique $c_{i,j} \in \mathbb{R}$ such that

$$[\Gamma_f] = \sum_{i,j} c_{i,j} \pi_1^\star[\omega_i] \cdot \pi_2^\star[\tau_j].$$

By definition, for every closed form η in $\Omega(M \times M)$ we have:

$$\int_M \iota^\star(\eta) = \int_{\Gamma_f} \eta = \int_{M \times M} \eta \wedge [\Gamma_f] = \sum_{i,j} c_{i,j} \int_{M \times M} \eta \cdot \pi_1^\star \omega_i \cdot \pi_2^\star \tau_j. \quad (*)$$

The idea is now simply to apply this to an η of the form $\pi_1^\star \tau_k \cdot \pi_2^\star \omega_l$, where k and l are arbitrary in I. The left hand side of the equation $(*)$ becomes:

$$\int_M \iota^\star(\pi_1^\star \tau_k \cdot \pi_2^\star \omega_l) = \int_M \tau_k \cdot f^\star \omega_l = \sum_m f_{m,l} \int_M \tau_k \omega_m$$

$$= \sum_m (-1)^{(\deg \omega_m)(\deg \tau_k)} f_{m,l} \delta_{m,k}$$

$$= (-1)^{(\deg \omega_k)(\deg \tau_k)} f_{k,l},$$

while the right hand side of $(*)$ becomes:

$$\sum_M c_{i,j} \int_{M \times M} \pi_1^\star \tau_k \cdot \pi_2^\star \omega_l \cdot \pi_1^\star \omega_i \cdot \pi_2^\star \tau_j$$

$$= \sum_{i,j} c_{i,j} (-1)^{(\deg \omega_l)(\deg \omega_i) + (\deg \tau_k)(\deg \omega_i)} \int_M \omega_i \wedge \tau_k \cdot \int_M \omega_l \wedge \tau_j$$

$$= \sum_{i,j} c_{i,j} (-1)^{(\deg \omega_l)(\deg \omega_i) + (\deg \tau_k)(\deg \omega_i)} \delta_{i,k}\, \delta_{j,l}$$

$$= c_{k,l} (-1)^{(\deg \omega_l)(\deg \omega_k) + (\deg \tau_k)(\deg \omega_k)}.$$

Combining the two we see that $c_{k,l} = (-1)^{\deg \omega_k} f_{k,l}$ (note that $f_{k,l} = 0$ if $\deg \omega_k \neq \deg \omega_l$). This completes the proof of the first identity for $[\Gamma_f]$.

The second identity follows from the definition of the $f_{i,j}$. The first expression for $[\Delta]$ follows from the one for $[\Gamma_{I_M}]$ because $I_M^\star \omega_i = \omega_i$. For the second one, we will use the fact that the transformation $(x, y) \mapsto (y, x)$ on $M \times M$ preserves the orientation on the diagonal Δ (as it induces the identity on it), but changes the orientation of $M \times M$ by a factor $(-1)^{\dim M}$: in local coordinates:

$$dy_1 \wedge \ldots \wedge dy_n \wedge dx_1 \wedge \ldots \wedge dx_n = (-1)^{n^2} dx_1 \wedge \ldots \wedge dx_n \wedge dy_1 \wedge \ldots \wedge dy_n,$$

and $n^2 - n = n(n-1)$ is always even. $\qquad\qquad\qquad\qquad\qquad\square$

Proposition 3.3.4.

$$\int_{M \times M} [\Gamma_f] \cdot [\Delta] = \sum_i (-1)^i \operatorname{Tr}\left((f^{\star,i}) : H^i(M) \to H^i(M)\right).$$

Proof. For convenience, set $l = \dim M + \deg \omega_i + \deg \omega_k$ and

$m = \dim M + \deg \omega_i + \deg \omega_k + \deg \tau_k \deg \omega_k + \deg \tau_k \deg \tau_j + \deg \omega_k \deg \tau_j.$

Then using Fubini's theorem,

$$\int_{M \times M} [\Gamma_f] \cdot [\Delta] = \sum_{i,j,k} (-1)^l f_{i,j} \int_{M \times M} \pi_1^\star \omega_i \pi_2^\star \tau_j \pi_2^\star \omega_k \pi_1^\star \tau_k$$

$$= \sum_{i,j,k} (-1)^m f_{i,j} \int_M \omega_i \wedge \tau_k \int_M \omega_k \wedge \tau_j$$

$$= \sum_{i,j,k} (-1)^m f_{i,j} \delta_{i,k} \delta_{k,j}$$

$$= \sum_i (-1)^{\dim M + \deg \tau_i} f_{i,i} = \sum_i (-1)^i \operatorname{Tr}(f^{\star,i} : H^i(M) \to H^i(M)).$$

$\square$

Completion of the proof of Theorem 3.3.2:

Proof. The result on transversal intersections tells us that

$$\int_{M \times M} [\Gamma_f] \cdot [\Delta] = \sum_{p=f(p)} \nu_p,$$

where $\nu_p = \pm 1$ depending on whether the two orientations on $T_{(p,p)}(M \times M) = N_{(p,p)}(\Gamma_f) \oplus N_{(p,p)}(\Delta)$ given by the orientations of $M \times M$, Γ_f and Δ do, or do not agree. We remind the reader that the direct sum decomposition of $T(M \times M)$ comes from $T_{(p,p)}(M \times M) = T_{(p,p)}(\Gamma_f) \oplus T_{(p,p)}(\Delta)$. So, we have to compare the given orientations on the first and last of the following vector spaces:

$$T_{(p,p)}(M \times M) = T_{(p,p)}(\Gamma_f) \oplus T_{(p,p)}(\Delta) = N_{(p,p)}(\Gamma_f) \oplus N_{(p,p)}(\Delta).$$

Let x be a fixed point and $(v_1, \ldots, v_n)$ be a positively oriented basis of $T_p(M)$. Then we have positive bases as follows:

1. $T_{(p,p)}(M \times M)$: $((v_1, 0), \ldots, (v_n, 0), (0, v_1), \ldots, (0, v_n))$.

2. $T_{(p,p)}(\Gamma_f)$: $(v_1, df_p(v_1)), \ldots, (v_n, df_p(v_n))$ (to be adjoined so as to give a positive orientation with $((0, v_1), \ldots, (0, v_n)))$.

3. $T_{(p,p)}(\Delta)$: $((v_1, v_1), \ldots, (v_n, v_n))$ (similarly completed with $((0, v_1), \ldots, (0, v_n)))$.

The completions of the positive bases of $T_{(p,p)}(\Gamma_f)$ and $T_{(p,p)}(\Delta)$ to positive bases of $T_{(p,p)}(M \times M)$ above show $((0, v_1), \ldots, (0, v_n))$ induces a positive basis of $N_{(p,p)}(\Gamma_f)$, and of $N_{(p,p)}(\Delta)$. Let us now consider the given isomorphism $T_{(p,p)}(\Delta) \to N_{(p,p)}(\Gamma_f)$ and compute the factor by which it changes the orientations. This is done by computing in $T_{(p,p)}(M \times M)/T_{(p,p)}(\Gamma_f)$. The element $(v_i, df_p(v_i))$ of $T_{(p,p)}(\Delta)$ is mapped to $\overline{(v_i, df_p(v_i))} = \overline{(0, v_i - df_p(v_i))}$. Hence the factor here for comparing with the basis of the $\overline{(0, v_i)}$ is sgn det$(I_M - df_p)$.

We do the same for $T_{(p,p)}(\Gamma_f) \to N_{(p,p)}(\Delta)$ by computing in $T_{(p,p)}(M \times M)/T_{(p,p)}(\Delta)$. The element $(v_i, df_p(v_i))$ is mapped to $(v_i, df_p(v_i)) = (0, df_p(v_i) - v_i)$. Hence the factor for comparing with the basis of the $(0, v_i)$ is sgn det$(df_p - I_M)$. Combining these, we see that the sign relating the orientations on $T_{(p,p)}(\Delta) \oplus T_{(p,p)}(\Gamma_f)$ and $N_{(p,p)}(\Gamma_f) \oplus N_{(p,p)}(\Delta)$ is the product of the signs of det$(I_M - df_p)$ and det$(df_p - I_M)$.

Finally, let us compute the sign between $T_{(p,p)}(M \times M)$ and $T_{(p,p)}(\Delta) \oplus T_{(p,p)}(\Gamma_f)$. We begin with our positive basis of the last term and apply elementary operations to it that preserve the orientation:

$$((v_1, v_1), \ldots, (v_n, v_n), (v_1, df_p(v_1)), \ldots, (v_n, df_p(v_n)))$$

$$\equiv ((v_1, v_1), \ldots, (v_n, v_n), (0, df_p(v_1) - v_1), \ldots, (0, df_p(v_n) - v_n))$$

$$\equiv ((v_1, 0), \ldots, (v_n, 0), (0, df_p(v_1) - v_1), \ldots, (0, df_p(v_n) - v_n)).$$

Hence the sign here is sgn $\det(df_p - I_M)$. This means the orientations on $T_{(p,p)}(M \times M)$ and $N_{(p,p)}(\Gamma_f) \oplus N_{(p,p)}(\Delta)$ are compatible up to the factor sgn $\det(I_M - df_p)$. $\qquad\qquad\qquad\qquad\qquad\qquad\qquad\qquad$ $\square$

3.4   Some Applications

Our first application of the Lefschetz theorem is to complex projective space (see [85]).

Consider the complex vector space $\mathbb{C}^{n+1}$, and let $A \in GL(n+1, \mathbb{C})$. We remind the reader that since a linear map A sends lines to lines, it induces a map $\tilde{A}$ on projective space $P(\mathbb{C}^n)$. A point $p \in P(\mathbb{C}^n)$ is a fixed point of $\tilde{A}$ if and only if it is an A-invariant line. Thus, fixed points of $\tilde{A}$ are the eigenvectors of A.

We now calculate $L(\tilde{A})$, for an $A \in GL(n+1, \mathbb{C})$ with distinct eigenvalues.

Proposition 3.4.1. *Such an A is a Lefschetz map on projective space $P(\mathbb{C}^n)$ and $L(\tilde{A}) = n + 1$.*

Proof. Since A has all eigenvalues λ_i, $i = 0, ..., n$ distinct, it is diagonalizable (over $\mathbb{C}$), i.e.

$$A = \begin{pmatrix} \lambda_0 & & & \\ & \lambda_1 & & \\ & & \ddots & \\ & & & \lambda_n \end{pmatrix}.$$

Since the fixed points of $\tilde{A}$ correspond to the $n + 1$ eigenvectors of A,

$$z_0 = \begin{pmatrix} 1 \\ 0 \\ \vdots \\ 0 \end{pmatrix}, \quad z_1 = \begin{pmatrix} 0 \\ 1 \\ \vdots \\ 0 \end{pmatrix}, \quad \cdots \quad, \quad z_n = \begin{pmatrix} 0 \\ 0 \\ \vdots \\ 1 \end{pmatrix}.$$

Let p_0 be the fixed point for $\tilde{A}$ corresponding to the eigenvector v_0. In homogeneous coordinates p_0 is the point $[1; 0; 0; ...; 0]$. Consider the chart U_0 around this point, i.e. the points $[1; z_1; z_2; ...; z_n]$. Then

$$\tilde{A}(p_0) = \begin{pmatrix} \lambda_0 & & & \\ & \lambda_1 & & \\ & & \ddots & \\ & & & \lambda_n \end{pmatrix} \cdot \begin{pmatrix} 1 \\ z_1 \\ \vdots \\ z_n \end{pmatrix} = \begin{pmatrix} \lambda_0 \\ \lambda_1 z_1 \\ \vdots \\ \lambda_n z_n \end{pmatrix} \sim \begin{pmatrix} 1 \\ \frac{\lambda_1}{\lambda_0} z_1 \\ \vdots \\ \frac{\lambda_n}{\lambda_0} z_n \end{pmatrix}.$$

This means in the neighborhood U_0 of p_0, $\tilde{A}$ is given by

$$\tilde{A}([1; z_1; ...; z_n]) = [1; \frac{\lambda_1}{\lambda_0} z_1; ...; \frac{\lambda_n}{\lambda_0} z_n].$$

As a linear map, its derivative $d\tilde{A}|_{p_0}$ is

$$d\tilde{A}|_{p_0} = \begin{pmatrix} \frac{\lambda_1}{\lambda_0} & & & \\ & \frac{\lambda_2}{\lambda_0} & & \\ & & \ddots & \\ & & & \frac{\lambda_n}{\lambda_0} \end{pmatrix}.$$

Since by hypothesis, all eigenvalues $\lambda_0,...,\lambda_n$ are distinct, all diagonal entries of the matrix $d\tilde{A}|_{p_0}$ are $\neq 1$, which implies $\tilde{A}$ is a Lefschetz map.

Now, we have to check the sign of $\det_{\mathbb{R}}(I - d\tilde{A}|_{p_0})$. To do this, we remark that $\mathrm{GL}(n, \mathbb{C})$ is connected (see e.g. [1], p. 9). Therefore it sits in $\mathrm{GL}^+(2n, \mathbb{R})$ (each complex coordinate splits in two real ones), therefore $\det_{\mathbb{R}}(I - d\tilde{A}|_{p_0}) > 0$, which implies the sign is $+1$. By repeating this calculation for all other fixed points, we find each gives us a $+1$. Thus the Lefschetz number $L(\tilde{A})$ is

$$L(\tilde{A}) = 1 + 1 + ... + 1 = n + 1.$$

Since $\mathrm{GL}(n + 1, \mathbb{C})$ is connected, $\tilde{A}$ is homotopic to the identity map, and the above result is consistent with the fact that the Euler characteristic of $P(\mathbb{C}^n)$ is $n + 1$.   □

3.4.1   Maximal Tori in Compact Lie Groups

An important application of the Lefschetz fixed point theorem is a basic structural result concerning compact connected Lie groups due to E. Cartan. Namely, the conjugacy of its maximal tori and the fact that the conjugates of any of them fill out the group (see e.g. Adams [2], the proof using fixed points is due to A. Weil). This result is important for the classification of the compact Lie groups and also for the classification of their irreducible representations.

Definition 3.4.2. When the compact connected Lie group is abelian it is called a *torus* T, Such groups are isomorphic to a finite product of circles. A *maximal torus* T of a compact connected Lie group G is a subgroup which is a torus not properly contained in any larger one.

Now, let G be a compact connected Lie group and X be vector in its Lie algebra $\mathfrak{g}$. By exponentiating X, we get a 1-parameter subgroup $H = \{e^{tX}, \ t \in \mathbb{R}\}$ of G which is connected and abelian. Therefore, its closure $\bar{H}$, is a compact, connected and abelian subgroup of G. In other words, a torus. Therefore by dimension maximal tori exist.

A torus $T \subset G$ (not necessarily maximal) acts on $\mathfrak{g}$ by the restriction of the adjoint representation $\mathrm{Ad} : G \to \mathrm{GL}(\mathfrak{g})$. Using Haar measure (see [1], p. 90) choose a positive definite symmetric form on $\mathfrak{g}$ which is bi-invariant under G (and hence Ad invariant). Then $\mathfrak{g}$ splits into orthogonal irreducible representations of T, which are of dimensions 1 and 2. By connectedness the ones of dimension 1 are trivial. For those of dimension of 2, choose an orthonormal basis and represent T by SO(2). Thus,

$$\mathfrak{g} = V_0 \oplus (\oplus_{i=1}^{k} V_i),$$

where T acts trivially on V_0 and acts on V_i (dim $V_i = 2$ for $i \geq 1$) by

$$\begin{pmatrix} \cos 2\pi\theta_i(t) & -\sin 2\pi\theta_i(t) \\ \sin 2\pi\theta_i(t) & \cos 2\pi\theta_i(t) \end{pmatrix},$$

where $\theta_i(t) : T \to \mathbb{R}/\mathbb{Z}$ as a nonzero map $\theta_i : \mathfrak{t} \to \mathbb{R}$ (here $\mathfrak{t}$ is the Lie algebra of T) taking integer values on the kernel of the (surjective) exponential map $\exp : \mathfrak{t} \to T$.

An important feature of tori is that their automorphism groups are discrete. Actually, any compact Lie group G has a discrete automorphism group. This is because here the topology on $\mathrm{Aut}(G)$ is that of *uniform convergence*. Let $W(G, U) = \{\alpha \in \mathrm{Aut}(G) : \alpha^{\pm 1}(G) \subseteq U\}$ be a neighborhood of I in $\mathrm{Aut}(G)$, where U is a neighborhood of 1 in G. Since G is a Lie group, we can choose U so that it contains no proper subgroups of G. Hence $W(G, U) = (I)$.

As we shall see in the case of a torus, T^n, $\mathrm{Aut}(G)$ can be calculated explicitly: $\mathrm{Aut}(T^n) = \mathrm{SL}^{\pm}(n, \mathbb{Z})$ (see Proposition 5.5.3).

Let G be a compact, connected Lie group. There is an action of a group (the Weyl group) on the space of all maximal tori T of G which we will now define.

Let $N(T)$ denote the *normalizer* of T, i.e.

$$N = N(T) = \{g \in G : gT = Tg\} = \{g \in G : gTg^{-1} = T\}$$

and $N(T)_0$ its identity component.

Using joint continuity of $\mathrm{Aut}(G) \times G \to G$, one sees easily that $N(T)$ is a closed subgroup of G, so $N(T)$ and $N(T)/N(T)_0$ are both compact. Moreover since $N(T)$ is a Lie group (since it is a closed subgroup of G), $N(T)_0$ is open in $N(T)$ so $N(T)/N(T)_0$ is discrete. Therefore $N(T)/N(T)_0$ is finite.

Proposition 3.4.3. $N(T)_0 = T$.

Proof. Evidently $N(T)_0 \supseteq T$. Consider the continuous map

$$\phi : N \to \mathrm{Aut}(T), \quad n \mapsto \phi(n)(t) = ntn^{-1}.$$

The map ϕ from the map $N \times T \to T$ is just the restriction of the conjugation map $G \times G \to G$. Since, as we have seen above, $\mathrm{Aut}(T)$ is discrete, $N(T)_0$ acts trivially on $\mathfrak{g}$. Therefore it centralizes G and so also T. Now, if $N(T)_0$ contains T properly, we can choose a 1-parameter subgroup H of $N(T)_0$ not contained in T. The subgroup $\langle T, H \rangle$, generated by H and T is connected and *abelian*. In addition, since $T \subseteq \langle T, H \rangle \subseteq \overline{\langle T, H \rangle} \subseteq N(T)_0$, $\langle T, H \rangle$ is compact. Hence, $\langle T, H \rangle$ is a torus, contradicting the maximality of T. Thus $N(T)_0 = T$.   □

Definition 3.4.4. The quotient group $W(G,T) := N(T)/T$ is called the *Weyl group* of G.

It is clearly finite by Proposition 3.4.3.

As an example, consider the group $G = \mathrm{U}(n)$ of unitary $n \times n$ matrices. Here (by the spectral theorem) a maximal torus T is the subgroup of all diagonal matrices,

$$\begin{pmatrix} e^{i\theta_1} & & & \\ & e^{i\theta_2} & & \\ & & \ddots & \\ & & & e^{i\theta_n} \end{pmatrix}$$

and the Weyl group is the symmetric group S_n, which acts on T by permuting the θ_i, $i = 1, ..., n$.

As we shall see, the fact that any unitary matrix can be diagonalized has a generalization valid for any compact connected Lie group called Cartan's maximal torus theorem.

Theorem 3.4.5. *Let G be a compact connected Lie group and T be a maximal torus, then*

$$G = \bigcup_{g \in G} gTg^{-1}.$$

Moreover, every maximal torus of G is conjugate to T.

Proof. First, we will prove that any element of G is conjugate to something in T; that is $G = \cup_{g \in G} g^{-1}Tg$. This means the coset xT is a fixed point of the action of g on G/T[4]. Denote this action by $f_g : G/T \to G/T$. That is, $f_g(xT) = gxT$. Thus, a fixed point xT of f_g is a coset such that $gxT = xT$. That is, $g \in xTx^{-1}$. So we seek a

---

[4]The space G/T has many interesting properties, for example it is a Kähler manifold, and even a projective variety because it is a flag manifold (see Chapter 6). For example, when $G = \mathrm{U}(n)$, G/T is the space of all complex flags in $\mathbb{C}^n$. Another important way of characterizing these spaces is as orbits of elements in the Lie algebra, $\mathfrak{g}$, under the adjoint action of G. Topologically, the cohomology ring of G/T turns out to be quite simple to describe, with all cohomology in even degree.

fixed point of f_g. Now, since G is connected, f_g is homotopic to f_e and therefore

$$L(f_g) = L(f_e) = L(I) = \sum_k (-1)^k \dim_{\mathbb{Q}}(G/T, \mathbb{Q}) = \chi(G/T).$$

By Lefschetz' theorem, to prove the existence of a fixed point, it is sufficient to show $\chi(G/T) \neq 0$. By the Kronecker approximation theorem (see [1], Appendix B) let g_0 be a generator of T, meaning the cyclic subgroup generated by it is dense, and consider the corresponding map (by translations) $f_{g_0} \equiv f_0$, which is homotopic to f_g. The fixed points of f_0 are exactly the cosets nT, where $n \in N_G(T)$. By Proposition 3.4.3 $T = N_G(T)_0$ has finitely many cosets. Therefore, f_0 has only finitely many fixed points and so these are isolated. To calculate $L(f_0)$ we have to calculate the degree $m_{f_0}(nT)$ for each $n \in N_G(T)$. We will show the degrees of all these fixed points (nT) are equal. To do so consider the map

$$r_n : G/T \to G/T, \quad gT \mapsto gnT.$$

This map sends the point T to the point nT, it is a diffeomorphism, and commutes with f_0, since

$$r_n(f_0(gT)) = r_n(g_0 gT) = ng_0 gT = g_0 ngT = f_0(r_n(gT)).$$

Hence,

$$(f_0)_{\star,nT} \circ (r_n)_{\star} = (r_n)_{\star} \circ (f_0)_{\star,T},$$

where the maps send the tangent space of G/T at the point T to the tangent space of G/T at the point nT. This implies

$$\det((f_0)_{\star,T} - I) = \det((f_0)_{\star,nT} - I),$$

which means the degrees of all fixed points are equal. So, we can just take the fixed point T and check whether its degree $\det((f_0)_{\star,T} - Id) \neq 0$, i.e. if f_0 is transversal. But from the discussion at the beginning of this section

$$
\mathrm{I} - (f_0)_{\star,T} = \left(
\begin{array}{cc|c}
\begin{array}{cc}
1 - \cos(2\pi\theta_1(g_0)) & \sin(2\pi\theta_1(g_0)) \\
- \sin(2\pi\theta_1(g_0)) & 1 - \cos(2\pi\theta_1(g_0))
\end{array} & 0 \\
\hline
0 & \ddots
\end{array}
\right)
$$

i.e. a 2×2-block diagonal matrix. Therefore the determinant is

$$
\det(\mathrm{I} - (f_0)_{\star,T}) = \prod_1^m \det \begin{pmatrix} 1 - \cos(2\pi\theta_i(g_0)) & \sin(2\pi\theta_i(g_0)) \\ - \sin(2\pi\theta_i(g_0)) & 1 - \cos(2\pi\theta_i(g_0)) \end{pmatrix}.
$$

Now, this is > 0 unless $\cos(2\pi\theta_k(g_0)) = 1$ for some k. Since g_0 is a generator of T, if $\cos(2\pi\theta_k(g_0)) = 1$, then $\theta_k(g_0) \equiv 0 \bmod 1$, contradiction, because θ_k is not trivial. Therefore the index of f_{g_0} at T is 1 and

$$
L(f_{g_0}) = |N(T)/T| \neq 0.
$$

Thus $L(f_g) \neq 0$ and so f_g has a fixed point.

Finally, to prove that any two maximal tori T, and T_1 are conjugate, let $t \in T$ be a generator of T. Then $t \in g^{-1}T_1 g$ for some $g \in G$. Thus $T \subset g^{-1}T_1 g$, and since T is maximal, $T = g^{-1}T_1 g$.   □

As a corollary we get,

Corollary 3.4.6. *For a compact connected Lie group, G, the exponential map is surjective.*

Proof. Let $g \in G$. By Theorem 3.4.5 g lies in some conjugate of T. Since exp commutes with conjugation we may as well assume $g \in T$ itself, where $\exp_T$ is surely surjective. The result follows since $\exp_G |_{\mathfrak{t}} = \exp_T$, where $\mathfrak{t}$ is the Lie algebra of T.   □

3.4.2   The Poincaré-Hopf's Index Theorem

We now turn to the important Poincaré-Hopf Index theorem. This was first proved for surfaces by Poincarè in 1885 and by H. Hopf in 1926 in general (see [58]) and was one of the early results linking a topological and a differential geometric invariant. (The earliest is perhaps the Gauss-Bonnet theorem.) H. Hopf's theorem asserts that if M is a C^∞ compact, connected manifold (without boundary) and X a smooth vector field on M with a finite number of zeros, the sum of the indices of the zeros of X is $\chi(M)$.

Its meaning is that the Euler characteristic of a compact manifold M is a measure of the obstruction to obtaining a smooth vector field without singularities. Since the singular points of a vector field are points of equilibrium of a dynamical system, it is not surprising that this result has many applications, from mathematical economics, optimization of communication systems, electrical engineering and applied probability, to electrodynamics, stability of molecular complexes in chemistry, crystallography, astrophysics, etc.... . For example, it explains why a fluid flowing smoothly on a 2-sphere must be stationary at some point, but on a torus need not have any stationary points.

To prove this result we first review some notions from Morse theory. Here we follow [70] and [48].

If $f : M \to \mathbb{R}$ is a smooth function, we call p in M a *critical point* of f, if $df_p : T_p(M) \to T_{f(p)}(\mathbb{R})$ is zero, i.e. in a local coordinate system $(U_p; x_1, ..., x_n)$ we have

$$\frac{\partial f}{\partial x_1}(p) = ... = \frac{\partial f}{\partial x_n}(p) = 0.$$

The number $f(p)$ is called a *critical value* of f.

Just as in the case of one variable, we look for *nondegenerate* critical points p, i.e. critical points with nonzero second derivative. That is,

$$\det \left(\frac{\partial^2 f}{\partial x_i \partial x_j} \right) \neq 0.$$

Definition 3.4.7. A function $f : M \to \mathbb{R}$ whose all critical points are nondegenerate is called a *Morse function*.

Lemma 3.4.8. (Morse Lemma). *If p is a nondegenerate critical point of $f : M \to \mathbb{R}$, then there is a local coordinate system $(x_1, x_2, ..., x_n)$ in a neighborhood U of p with $x_1(p) = x_2(p) = ... = x_n(p) = 0$ such that,*

$$f(x) = f(p) - x_1^2 - x_2^2 - \cdots - x_k^2 + x_{k+1}^2 + \cdots + x_n^2$$

for any $x \in U$. Furthermore, any such coordinate system will give the same number of positive and negative terms.

For a proof see [79], p. 218. From this it follows that nondegenerate critical points are isolated.

Definition 3.4.9. The number of negative terms in the above expression is called the *Morse index* of f at p. (This measures the number of independent directions in which f is locally decreasing.)

The index of a vector field at an isolated singularity can be defined in various ways, but we will limit ourself to the classical ones. Since we are interested in compact manifolds M, as before we may assume $M \subset \mathbb{R}^n$ for some n A vector field X on M is a smooth map X on M where at each point $X(p) \in T_p(M)$.

Now, let p be an isolated zero of X. To define the (integer-valued) *index* $\mathrm{Ind}_p(X)$ of X at p, first, we assume that M is an open domain in $\mathbb{R}^n$. Let $B_\epsilon(p)$ be a small ball around p so that there is no other zero of X within $B_\epsilon(p)$. We define the *Gauss map* by,

$$\gamma : \partial B_\epsilon(p) \equiv S^{n-1} \to S^{n-1} \quad : x \mapsto \gamma(x) := \frac{X(x)}{\| X(x) \|}.$$

Definition 3.4.10. The (local) *index* of X at p, denoted by $\mathrm{Ind}_p(X)$, is the degree of the Gauss map $\gamma : S^{n-1} \to S^{n-1}$.

One can prove that the local index does not depend on the choice of the $B_\epsilon(p)$, the choice of coordinates, or on the orientation.

Let X be a vector field, $p \in M$ be a point. If $X(p) = 0$ we shall call p a *critical point* of X. Let p be a critical point of X. We say p is *nondegenerate* critical point of X if the derivative, $\nabla X_p : T_p(M) \to T_p(M)$ is invertible. In this case

$$\mathrm{Ind}_p(X) = \mathrm{sgn}(\det(\nabla X_p)) \in \{\pm 1\}.$$

Lemma 3.4.11. *On any compact manifold M, there is a vector field X with only nondegenerate critical points p such that*

$$\sum_{p:\, X(p)=0} \text{Ind}_p(X) = \chi(M).$$

Proof. By Morse Lemma (see Lemma 3.4.8) we can find a map $f : M \to \mathbb{R}$ such that if $(x_1, ..., x_n)$ are Morse coordinates at the critical point $p = (0, 0, ..., 0)$ in M. Then

$$f(x_1, ..., x_n) = f(p) - x_1^2 - \cdots - x_k^2 + x_{k+1}^2 + \cdots + x_n^2,$$

where k is the index of p. By definition, p is a critical point if $df|_p = 0$, i.e. $\nabla f|_p = 0$. Hence p is also a zero of the vector field defined by the gradient of f. Setting $X = \frac{1}{2}\nabla f$ we get

$$X(x) = (-x_1, -x_2, \ldots, -x_k, x_{k+1}, ..., x_n)$$

which means that X can be seen as the product of two vector fields, Y on $\mathbb{R}^k$, with $Y(y) = -y$, and Z on $\mathbb{R}^{n-k}$ with $Z(z) = z$. Since $\text{Ind}_p(Y) = (-1)^k$ and $\text{Ind}_p(Z) = 1$, we get

$$\text{Ind}_p(X) = (-1)^k.$$

Therefore, the sum of indices of zeros of X is $\sum_0^n (-1)^k b_k$, where b_k is the number of critical points of f of index k. This is equal to $\chi(M)$. $\square$

In Section 3.3.2 we saw what transversality means. Now we note that transversality is a property stable under small perturbations. That is, submanifolds (or maps) which are transversal remains so under small perturbations. Actually, more is true. Using Sard's theorem (see [79], p. 283), one can show that not only is transversality stable, it is *generic*, where here generic means that even if a statement is not true, it can be made true by a small perturbation.

Indeed, we have the following result (see [48], p. 68): Suppose $f_t : M \to N$ is a family of smooth maps indexed by $t \in I = [0, 1]$. Consider the map $F : M \times I \to N$ defined by $F(p, t) := f_t(p)$ which we also assume to be smooth. (Here we allow M to have a boundary.)

Theorem 3.4.12. (Transversality Theorem.) *Let M and $F : M \times I \to N$ be as above. If L is any boundaryless submanifold of N and both F and ∂F are transversal to L, then for almost every $t \in I$ (Lebesgue measure), f_t and ∂f_t are transversal to L.*

In particular, by regularity this occurs on a dense subset of I.

Now, we are ready for the Poincaré-Hopf theorem. We present the proof given by Hans Samelson in [89]. In this proof we will need the notion for the zero section of the tangent bundle $T(M)$. This is the vector field,

$$X = \{(p, \xi) \in T(M), \ \xi = 0 \text{ at } T_p(M)\}.$$

Its significance is that this vector field is actually the manifold M in disguise (given by $p \mapsto (p, 0_p)$, where $0_p \in T_p(M)$ for each $p \in M$).

Theorem 3.4.13. (Poincaré-Hopf Theorem.) *Let M be a smooth compact, connected, manifold, and X be a C^∞ vector field on M with a finite number of zeros. Then, the sum of the indices of the zeros of X is $\chi(M)$.*

Proof. The proof proceeds in two stages. First, we show that if X and Y are C^∞ vector fields on M with a finite number of zeros, that the sum of the indices of their zeros is the same for each. The second step consists of considering the "gradient field" of a Morse function on M and showing this common value is $\chi(M)$.

From the comment on transversality just after the theorem, if we slightly deform X to obtain X_1, then any zero of X_1 will be near a zero of X, and the index of any zero of X will be the sum of the indices of the nearby zeros of X_1. Using the transversality theorem, we can restrict ourselves to transversal vector fields. Now X has only isolated zeros. In addition, each index is $+1$ or -1, because when we consider X as a submanifold of $T(M)$, the tangent space X_p to X at a zero, p, can be regarded as the graph of a linear map from M_p, the (horizontal) tangent space of the 0-section M of $T(M)$ at p to M_p, the (vertical) tangent space to the fiber M_p of $T(M)$ at 0. Transversality implies this map $M_p \to M_p$ is nonsingular. Therefore, since the index is the sign of the determinant, it must be $+1$, or -1 according to the preservation of the orientation, or not.

Now suppose X and Y are transversal vector fields. We consider the manifold $T(M) \times I$, with $I = [0, 1]$. This is a vector bundle over $M \times I$, with 0-section $M \times I$ and fiber M_p at (p, t). The fields X and Y give a section of this bundle over the subset $M \times \{0\} \cup M \times \{1\}$ of $M \times I$. We extend this to a continuous section W over $M \times I$, in other words set

$$W(p, t) = \begin{cases} (1 - 3t)X(p) & 0 \leq t \leq \frac{1}{3}, \\ 0 & \frac{1}{3} \leq t \leq \frac{2}{3}, \\ (3t - 2)Y(p) & \frac{2}{3} \leq t \leq 1. \end{cases}$$

By the transversality theorem, approximate W by a smooth section Z, which is transversal to $M \times I$ and such that $W = Z$ on

$$M \times [0, \epsilon] \cup M \times [1 - \epsilon, 1]$$

for some $\epsilon > 0$, where W is already smooth and transversal. We get

$$Z = \begin{cases} X & \text{on } M \times \{0\} \\ Y & \text{on } M \times \{1\}. \end{cases}$$

By the main property of transversality the intersection of Z (regarded as a submanifold of $T(M) \times I$) and $M \times I$ is of dimension 1 and consists of a finite number of disjoint closed curves (in $M \times (0, 1)$) and arcs (whose endpoints are on $M \times \{0\} \cup M \times \{1\}$). The endpoints of the arcs are precisely the zeros of X and Y.

Now, suppose that such an arc γ has both endpoints, $(a, 0)$ and $(b, 0)$, on $M \times \{0\}$. Then moving along γ, one of the indices of X at a and b is $+1$ and the other is -1. First we note that the tangent space to $T(M) \times I$ at a point (p, t) of $M \times I$ is of the form

$$M_p(\text{horizontal}) \oplus M_p(\text{vertical}) \oplus \mathbb{R}.$$

Hence it has a canonical orientation. Take a frame at $(a, 0)$, consisting of $\dot{\gamma}$ (tangent to γ), horizontal vectors $X_1, ..., X_n$ (a basis for M_a), and vertical vectors $Y_1, ..., Y_n$, images of the X_i under the linear map whose graph is the tangent space to X, described above. We move this frame continuously along γ, always with $\dot{\gamma}$ as first vector, and always using the tangent space to Z to map the horizontal X_i to vertical Y_i, the X_i at $(b, 0)$ should be in M_b. The fact that $\dot{\gamma}$ points to $M \times I$ at $(a, 0)$, but out of $M \times I$ at $(b, 0)$, means that the orientations of $M_a \oplus M_a$ and $M_b \oplus M_b$, defined by $X_1, ..., X_n, Y_1, ..., Y_n$, are opposite to one other. But that is equivalent to our claim about the indices of X at a and b. Similarly for arcs with both ends on $M \times \{1\}$.

Finally, if an arc goes across from $M \times \{0\}$ to $M \times I$, using the same argument we see the two endpoints have the same index, both $+1$ or both -1. It is clear now that the sum of ± 1's over the zeros of X equals that over the zeros of Y (the contributions of the endpoints of any arc that does not go "across" cancel).

To complete the proof consider the vector field $X = \frac{1}{2}\nabla f$, where f is the Morse function of Lemma 3.4.11. This gives us as the common value of the sums of indices of vector fields on M, namely $\chi(M)$. $\square$

As a simple application of the tubular neighborhood theorem we have:

Proposition 3.4.14. *If M is a compact manifold which admits a continuous nonzero tangent vector field, then there exists $f : M \to M$ without fixed points, homotopic to the identity.*

Proof. We first imbed M in $\mathbb{R}^n$ for some n and then work in the metric space $\mathbb{R}^n$. Let $X : M \to TM$ be the nonzero vector field. Since M is compact, for all $\epsilon > 0$ there is a $c > 0$ such that for each x the vector field $cX(x)$ has length less than ϵ. The function $F(t, x) = \pi(x + tcX(x))$ has no fixed point at $t = 1$. For if it had such a point, $\pi(x + cX(x)) = x$, then $X(x)$ would be both tangent and normal to M, which is impossible. Thus F defines a homotopy from f to the identity on M. $\square$

This last proposition can now be used to prove a simplified version of the Hopf index theorem (which is also called Hopf's theorem).

Corollary 3.4.15. *Let M be a compact smooth manifold, if $\chi(M) \neq 0$, then M does not admit a nonzero continuous tangent vector field.*

Proof. It suffices to observe that $\chi(M) = L(1_M)$. If M had a nonzero continuous tangent vector field, the previous proposition would give us a function $f : M \to M$ without fixed points homotopic to the identity. Hence $L(f) = \chi(M) \neq 0$, contradicting the Lefschetz theorem.   □

As a final corollary we recapture a previous result.

Corollary 3.4.16. *An even dimensional sphere, S^n, cannot admit a nonzero continuous tangent vector field.*

Proof. Here

$$H^k(S^n, \mathbb{Q}) = \begin{cases} \mathbb{Q} & k = n, \\ 0 & k \neq n \end{cases}.$$

Therefore, $L(f) = \chi(S^n) = 1 + (-1)^n = 2 \neq 0$.   □

From the proof we see odd dimensional spheres must be quite different. Indeed, it is well known that $M = S^1$, S^3 or S^7 are parallelizable. That is, each has smooth tangential vector fields $X_1, \dots, X_n$, where $n = \dim M$, which are linearly independent at every T_p, $p \in M$. The reason for this is that the first two of these are Lie groups while the third is almost a Lie group (except that multiplication is not associative) and Lie groups are always parallelizable (using left translation).

3.5   The Atiyah-Bott Fixed Point Theorem

In this section we give an introduction to the Atiyah-Bott fixed point theorem which should more properly be called the Atiyah-Bott fixed point formula. This is a significant generalization of the Lefschetz fixed point formula. Although we will not prove this important result, by applying what we do prove in the special case of De Rham cohomology we are able to recapture the Lefschetz fixed point theorem 3.3.2.

First a bit of history. In 1964, Atiyah and Bott participated at an algebraic geometry conference at Woods Hole. In a interview in the AMS Notices (vol. 48, no. 4, April 2001, 374-382) Bott recounted:

"....By that time, we (Atiyah and Bott) had learned to define an elliptic complex, and we saw the old de Rham theory in a new light: namely, that it satisfied the natural extension to vector bundles of the classical notion of ellipticity for a system of PDEs. During that conference we discovered our fixed point theorem, the Lefschetz fixed point theorem in this new context. This was a very pleasant insight. The number theorists at first told us we must be wrong, but then we turned out to be right. So we enjoyed that!

In a way, I always thought of the Lefschetz theorem as a natural first step on the way to the index theorem. You see, in the index theorem you compute the Lefschetz number of the identity map. The identity map has a very large fixed point set. So, if you have an idea that the Lefschetz number has to do with fixed points, then of course it is much easier to first try to prove the Lefschetz theorem for a lower-dimensional fixed point set. The fixed point theorem we proved in Woods Hole dealt precisely with the case in which the fixed point set was zero dimensional. Over the years I've encouraged people to study it over bigger and bigger fixed point sets and approach the final answer in this way. The analysis needed for the Lefschetz theorem in the case that we studied is very simple compared to the analysis needed for the true index theorem. Nevertheless, this special case fit in nicely with many things, and we could use it to prove some theorems about actions of finite groups on spheres and so on...."

Atiyah also writes about their discovery in *A Personal History* (published in *The Founders of Index Theory*, Edited by S. T. Yau, International Press, 2003):

"... One important refinement of the index theorem was the Lefschetz fixed point theorem for elliptic operators. This came in two models. First, for isolated fixed points, there were my two papers with Bott. These arose from a conjecture of Shimura which was drawn to our attention at the Woods Hole conference in 1964. Despite a dispute with the local experts on elliptic curves we soon became firmly convinced of the validity of the general formula, particularly once we realized that it included the famous Weyl character formula for compact Lie groups. Getting the proof was not too difficult, but our first attempt involved learning about the zeta functions of elliptic operators, which have subsequently been put to extensive use. The second model of the fixed point formula related to fixed point sets of any dimension, but only for the case of a compact Lie group action...."

Our use of the Lefschetz number dealt with the cotangent bundle of a *compact* manifold. Atiyah and Bott replaced this bundle by an entire family of bundles and they defined the corresponding cohomology which we now describe. We begin by reminding the reader of the notion of a real (or complex) vector bundle over a manifold M. Here $k = \mathbb{R}$ or $\mathbb{C}$.

Definition 3.5.1. A (*differentiable*) *k-vector bundle of rank* r over a manifold M is a quadruple (E, M, π, V) with the following properties:

- E, M are smooth manifolds and V is a r-dimensional k-vector space.

- $\pi : E \to M$ is a surjective submersion. We set $E_p := \pi^{-1}(p)$ and we will call it the *fiber* (of the bundle) over p.

- There exists a *trivializing cover*, i.e. an open cover $\mathcal{U} = (U_\alpha)_{\alpha \in A}$ of M and diffeomorphisms

$$\psi_\alpha : E \mid U_\alpha = \pi^{-1}(U_\alpha) \to V \times U_\alpha$$

with the following properties:

- For every $\alpha \in A$ the diagram below is commutative.

$$
\begin{array}{ccc}
E \mid U_\alpha & \xrightarrow{\psi_\alpha} & V \times U_\alpha \\
\pi \downarrow & & \downarrow proj \\
U_\alpha & = & U_\alpha
\end{array}
$$

- For every $\alpha, \beta \in A$ there exists a smooth map

$$
g_{\beta\alpha} : U := U_\alpha \cap U_\beta \longrightarrow \mathrm{Aut}(V), \quad u \mapsto g_{\beta\alpha}(u)
$$

such that for every $u \in U_{\alpha\beta}$ we have the commutative diagram

$$
\begin{array}{ccc}
V \times \{u\} & \xrightarrow{g_{\beta\alpha}(u)} & V \times \{u\} \\
\psi_\alpha|E_u \uparrow & & \uparrow \psi_\beta|E_u \\
E_u & = & E_u
\end{array}
$$

M is called *the base*, E is called *the total space*, V is called the *model (standard) fiber* and π is called the *canonical* (or *natural*) *projection*. A k-vector bundle of rank 1 is called a *line bundle*.

Definition 3.5.2. If $\pi : E \to M$ is a k-vector bundle over M, we call a *section* of E a map $s : M \to E$ such that $\pi \circ s = I_M$. The set of all sections of E is denoted by $\Gamma(E)$.

Let M be a manifold and let $\pi : E \to M$ be a complex vector bundle over M.

Definition 3.5.3. A *Hermitian metric* h on E is an assignment of a Hermitian inner product $\langle,\rangle_p$ to each fiber E_p of E such that for any open set $U \subset M$ and any two sections $\zeta, \eta \in \Gamma(U, E)$, the function

$$
\langle \zeta, \eta \rangle : U \to \mathbb{C}, \quad p \mapsto \langle \zeta(p), \eta(p) \rangle_p
$$

is differentiable. A complex vector bundle E equipped with a Hermitian metric h is called a *Hermitian vector bundle*.

We now prove,

Proposition 3.5.4. *Every complex vector bundle* $\pi : E \to M$ *admits a Hermitian metric.*

Proof. Let U_α be a locally finite cover of M with local frames $e_1^a, ..., e_r^a$ for E, we can define a Hermitian metric on $E|_{U_\alpha}$ by setting for any $\zeta, \eta \in \Gamma(U, E)$

$$\langle \zeta, \eta \rangle_p^\alpha := \sum_i \zeta_i \bar{\eta}_i,$$

and, if we denote by ϕ_α a smooth partition of unity subordinate to the cover, then

$$\langle \zeta, \eta \rangle_p := \sum_\alpha \phi_\alpha \langle \zeta, \eta \rangle_p^\alpha,$$

is a Hermitian inner product on E_p, and the map $p \mapsto \langle \zeta, \eta \rangle_p$ is smooth. $\square$

We now define morphisms of vector bundles.

Suppose (E, π_E, M, V) and (F, π_F, M, W) are two smooth k-vector bundles over M of rank r and q respectively. Assume $\{U_\alpha, \psi_\alpha\}_\alpha$ is a trivializing cover for E and $\{V_\beta, \phi_\beta\}_\beta$ is a trivializing cover for F. Then,

Definition 3.5.5. A *vector bundle morphism* from E to F is a smooth map $T : E \to F$ satisfying the following conditions:

- The following diagram is commutative.

$$
\begin{array}{ccc}
E & \xrightarrow{\ \ T\ \ } & F \\
\pi_E \downarrow & & \downarrow \pi_F \\
M & =\!=\!= & M
\end{array}
$$

- T is linear along the fibers, i.e. for every $p \in M$ and every α and β such that $p \in U_\alpha \cap V_\beta$ the composition, $\phi_\beta \circ T \circ \psi_\alpha^{-1}|_{E_p} : V \to W$

is linear, i.e.,

$$E_p \xleftarrow{\psi_\alpha^{-1}|E_p} V \times \{p\}$$

$$T|E_p \Big\downarrow \qquad\qquad \Big\downarrow \text{linear}$$

$$F_p \xrightarrow[\phi_\beta|F_p]{} W \times \{p\}$$

T is called an *isomorphism* if it is a diffeomorphism. A *gauge transformation* of E is a bundle automorphism $E \to E$.

This leads to,

Definition 3.5.6. A *subbundle* of E over M is a smooth submanifold $F \hookrightarrow E$ with the property that $\pi : F \to M$ is a vector bundle and the inclusion $F \hookrightarrow E$ is a bundle morphism.

We now define the pull back or induced bundle. Let N and M be smooth manifolds and $f : N \to M$ a smooth map. If (E, π, M) is a k-vector bundle over M then,

Definition 3.5.7. We define the *pullback* or *induced* bundle $f^\star E$ (or equivalently $f^{-1}E$) the bundle $(f^\star E, \tilde{\pi}, N)$ such that the following diagram is commutative:

$$f^\star E \xrightarrow{\ f^\star\ } E$$

$$\tilde{\pi} \Big\downarrow \qquad\qquad \Big\downarrow \pi$$

$$N \xrightarrow[\ f\ ]{} M$$

Here,
$$f^\star E = \{(n, e) \in N \times E \mid f(n) = \pi(e)\}$$
and $\tilde{\pi}(n, e) = n$, and $f^\star(n, e) = e$.

If $\psi_\alpha : E|U_\alpha \to U_\alpha \times V$ is a trivialization of E in a neighborhood of $f(n)$, then the map

$$f^\star\psi_\alpha : f^\star E_{f^{-1}U_\alpha} \to f^\star U_\alpha \times V$$

gives $f^\star E$ its manifold structure over the open set $f^{-1}(U)$. The transition functions for the pullback $f^\star E$ will be the pullback of the transition functions for E.

One sees easily that $(f \circ g)^\star E \cong g^\star \circ f^\star E$, and in the special case where f is the inclusion map $i : N \hookrightarrow M$ we recover $E|_N \cong i^\star E$.

Concerning sections, a map $f : M \to M$ induces a natural map

$$\Gamma_f : \Gamma(E) \longrightarrow \Gamma(f^{-1}E),$$

by composition: $\Gamma_f(s) = s \circ f$. Although there is no natural way to induce a map of sections: $\Gamma(E) \longrightarrow \Gamma(E)$, if there is a bundle map $\phi : f^{-1}E \longrightarrow E$, then the composition

$$\Gamma(E) \xrightarrow{\ \Gamma_f\ } \Gamma(f^{-1}E) \xrightarrow{\ \tilde{\phi}\ } \Gamma(E)$$

is an endomorphism of $\Gamma(E)$. Any such bundle map $\phi : f^{-1}E \longrightarrow E$ is called a *lifting* of f to E. At each point $x \in M$, a lifting ϕ is just a linear map of the fibers, $\phi_x : E_{f(x)} \to E_x$.

In the case of the de Rham complex, a map $f : M \to M$ induces a linear map $f_p^\star : T_{f(p)}^\star(M) \to T_p^\star(M)$ and hence a linear map

$$\bigwedge^k f_p^\star : \bigwedge^k T_{f(p)}^\star(M) \longrightarrow \bigwedge^k T_p^\star(M),$$

which is the lifting that finally defines the pullback of differential forms $f^\star : \Gamma(\bigwedge^k T^\star(M)) \longrightarrow \Gamma(\bigwedge^k T^\star(M))$.

Now, suppose M is an n-dimensional compact Riemaniann manifold and E and F are two k-vector bundles over M. Let $\Gamma(E)$ and $\Gamma(F)$ be their respective spaces of sections. We now define a differential operator between them.

Definition 3.5.8. A linear operator $D : \Gamma(E) \to \Gamma(F)$ is a *differential operator of order* d if, in any local trivializations of E and F over a coordinate chart $(U; x_1, ..., x_n)$, one has

$$Du(x) = \sum_{|\alpha| \le d} a^\alpha(x) \partial_\alpha u(x),$$

where $\alpha = (\alpha_1, ..., \alpha_m)$ is a multi-index with each $\alpha_i \in \{1, ..., n\}$, $|\alpha| = m$, $\partial_\alpha = \partial_{\alpha_1} \cdot \partial_{\alpha_2} \cdots \partial_{\alpha_m}$, and $a^\alpha(x)$ is the matrix representing an element of $\mathrm{Hom}_k(E_x, F_x)$, $x \in M$.

For a differential operator D of order d, we define its symbol[5]:

Definition 3.5.9. The *symbol,* $\sigma(D, \xi)$ of D is defined for any $p \in M$ and $\xi \in T_p^\star(M)$ by taking only the terms of order d in D:

$$\sigma(D, \xi)(x) = \sum_{|\alpha|=d} a^\alpha(x)\xi_\alpha,$$

where $\xi = \xi_1 dx_1 + \xi_2 dx_2 + \cdots + \xi_n dx_n$, and $\xi_\alpha = \xi_{\alpha_1}\xi_{\alpha_2}\cdots\xi_{\alpha_d}$. It is a homogeneous polynomial in the variable ξ of degree d with values in $\mathrm{Hom}(E_x, F_x)$.

In other words, the symbol of D is obtained by first discarding all but the highest-order terms of D and then replacing $\partial/\partial x^\alpha$ by ξ_α.

An equivalent definition of the symbol is the following which is very useful in calculating it: Let $E \to M$ be a vector bundle, $p \in M$, $\xi \in T_p^\star(M)$ and $s \in \pi^{-1}(p) = E_p$. Now, consider a section $\tilde{s} \in \Gamma(M, E)$ with $\tilde{s}(p) = s$ and a function $f \in C^\infty(M)$ such that $f(p) = 0$ and $df|_p = \xi \in T_p^\star(M)$. The symbol $\sigma(D, \xi)$ of the differential operator D can be defined as:

$$\sigma(D, \xi)(s) = \frac{1}{n!} D(f^n \tilde{s})|_p.$$

We observe that since $f(p) = 0$, the factor f^n automatically conserves the nth order term.

---

[5]We recall that a map $f : V \to W$ between two vector spaces is called *polynomial of degree m* if in some basis $\{e_1, e_2, \ldots, e_n\}$ of V it is given by

$$f\left(\sum_{i=1}^n x_i e_i\right) = \sum_{|\alpha|=m} x_1^{\alpha_1} x_2^{\alpha_2} \cdots x_n^{\alpha_n} w_a,$$

with $w_\alpha \in W$. The symbol of a differential operator will be a smooth section of the bundle $P^m(T^\star(M), \mathrm{Hom}_k(E_1, E_2))$, i.e. for each $x \in M$ we get a polynomial mapping of degree m on $T_x^\star(M)$ with values in the space of linear maps $E_x^1 \to E_x^2$.

Definition 3.5.10. A differential operator $D : \Gamma(E) \to \Gamma(F)$ is called *elliptic* if for any $p \in M$ and $\xi \neq 0$ in $T_p^\star(M)$, the symbol $\sigma(D, \xi)(p) :$ $E_p \to F_p$ is injective.

As an example we illustrate these concepts with the *Laplacian*.

$$\triangle = \frac{\partial^2}{\partial x_1^2} + \cdots + \frac{\partial^2}{\partial x_n^2}.$$

Its symbol is

$$\sigma(\triangle, \xi) = \sum_j (\xi_j)^2.$$

To see this, we calculate:

$$\sigma(\triangle, \xi)s = \frac{1}{2} \triangle (f^2 \tilde{s})|_x = \frac{1}{2} \sum_j \frac{\partial^2}{\partial x_j^2}(f^2 \tilde{s})|_x =$$

$$\frac{1}{2}\left(f^2 \triangle s \triangle \tilde{s} + 2f \, \triangle f \tilde{s} + 2f \sum_j \frac{\partial f}{\partial x_j} \frac{\partial \tilde{s}}{\partial x_j} + 2 \sum_j \frac{\partial f}{\partial x_j} \frac{\partial f}{\partial x_j} \tilde{s} \right)|_x = \sum_j (\xi_j)^2 s.$$

Now, if $\xi \neq 0$, the symbol is invertible, hence the Laplacian $\triangle$ is an elliptic operator.

Other examples can be found in M. Nakahara [81].

We now define the cohomology of an elliptic complex. Let M be a compact n-dimensional Riemannian manifold without boundary.

Definition 3.5.11. The differential complex

$$0 \longrightarrow \Gamma(E_0) \xrightarrow{\ D\ } \Gamma(E_1) \xrightarrow{\ D\ } \cdots \xrightarrow{\ D\ } \Gamma(E_k) \longrightarrow 0 \qquad (*)$$

is called *elliptic* if $D^2 = 0$ and for each $p \in M$ and any nonzero $\xi \in T_p^\star(M)$, the sequence

$$0 \longrightarrow E_{0,p} \xrightarrow{\sigma(D,\xi)} E_{1,p} \xrightarrow{\sigma(D,\xi)} \cdots \xrightarrow{\sigma(D,\xi)} E_{k,p} \longrightarrow 0$$

is exact.

Definition 3.5.12. We define the ith *cohomology group of the elliptic complex* $(E = (E_i), D = (D_i))$ by

$$H^i(E, D) = \operatorname{Ker} D_i / \operatorname{Im}(D_{i-l}).$$

We remark that because of the ellipticity of the complex and the compactness of M it follows that, for each i, $\dim H^i(E, d) < \infty$.

Now for the definition of the "Lefschetz number" of such a complex, we take as a *geometric endomorphism* $T := T(f, \phi)$ of the complex $(*)$ to be a collection of linear maps $T_i : \Gamma(E_i) \to \Gamma(E_i)$ such that

$$T_{i+1} \circ D = D \circ T_i,$$

for all i. Such a collection $T = \{T_i\}$ induces maps in cohomology

$$T_i^\star : H^i(E) \to H^i(E).$$

The *Lefschetz number* of T is defined to be,

$$L(T) = \sum (-1)^i \operatorname{Tr}(T_i^\star).$$

All this leads to the Atiyah-Bott fixed point theorem which calculates the Lefschetz number of a self map f of a compact manifold M.

Theorem 3.5.13. (Atiyah-Bott Fixed Point Theorem.) *Suppose we are given an elliptic complex* $(*)$ *on a compact manifold* M. *If for each i, the map $f : M \to M$ has a lifting $\phi_i : f^{-1}E_i \to E_i$ such that the induced maps $T_i : \Gamma(E_i) \to \Gamma(E_i)$ give a geometric endomorphism T of the elliptic complex, then the Lefschetz number of T is given by*

$$L(T) = \sum_{f(p)=p} \frac{\sum (-1)^i \operatorname{Tr} \phi_{i,p}}{|\det(1 - f_{\star,p})|}.$$

For a proof[6] see [6], and [7].

---

[6] This important result has a number of different proofs by various authors. Among them, Berline and Vergne, Donnelly, Gilkey, Kotake, Lafferty, Patodi, and several others.

3.5.1   The Case of the de Rham Complex

Elliptic complexes are quite common in differential geometry. A basic example is the de Rham complex. Here, we apply the Atiyah-Bott theorem, in the case of the de Rham complex to recover another proof of the Lefschetz Fixed Point Theorem for compact manifolds, thus showing the Atiyah-Bott formula is a generalization of the Lefschetz formula. Though the de Rham complex is naturally a real complex, we will work with its complexification. The reason for this will become apparent shortly.

Let M be a compact n-dimensional manifold and consider

$$\Omega^r = \Gamma\left(\bigwedge^r (T^\star(M)) \otimes_{\mathbb{R}} \mathbb{C} \right).$$

The usual exterior derivative operator $d : \bigwedge^r \to \bigwedge^{r+1}$ is a first order differential operator and we have the de Rham complex,

$$0 \longrightarrow \Omega^0 \xrightarrow{\; d \;} \Omega^1 \xrightarrow{\; d \;} \cdots \xrightarrow{\; d \;} \Omega^n \longrightarrow 0.$$

As is well known $d^2 = 0$, so this complex is elliptic.

We now compute the symbol of d.

Proposition 3.5.14. *The symbol of d is $\sigma(d, \xi) = \xi \wedge -$.*

Proof. Let $p \in M$, $\omega \in \Omega^r_p(M) \otimes \mathbb{C}$, $f \in C^\infty(M)$ with $f(p) = 0$, $df|p = \xi$, $\tilde{s} \in \Omega^r(M) \otimes \mathbb{C}$ and $\tilde{s}(p) = \omega$. To find the symbol for d we just note that

$$\sigma(d, \xi)\omega = d(f\tilde{s})|_p = df \wedge \tilde{s} + f d\tilde{s}|_p = \xi \wedge \omega,$$

since $fd\tilde{s}|_p = fd(d\omega) = 0$. Therefore

$$\sigma(d, \xi) = \xi \wedge -$$

in other words the symbol

$$\sigma(d, \xi) : \bigwedge^r T_p^\star(M) \otimes \mathbb{C} \longrightarrow \bigwedge^{r+1} T_p^\star(M) \otimes \mathbb{C},$$

is exterior multiplication by ξ. $\square$

We need the following proposition:

Proposition 3.5.15. *If V is an n-dimensional real or complex vector space, and $\xi \in V$, then the following complex*

$$0 \longrightarrow \textstyle\bigwedge^0 V \xrightarrow{\xi \wedge -} \textstyle\bigwedge^1 V \xrightarrow{\xi \wedge -} \cdots \xrightarrow{\xi \wedge -} \textstyle\bigwedge^n V \longrightarrow 0$$

is exact if and only if $\xi \neq 0$.

Proof. If $\xi = 0$ there is nothing to prove. Suppose $\xi \neq 0$ and assume that $\{\xi, e_2, ..., e_n\}$ is a basis of V. Then, relative to this basis, any $\eta \in \bigwedge^k V$ can be written as

$$\eta = \textstyle\sum_{1 < i_2 < ... < i_k \leq n} c_{i_2...i_k} \; \xi \wedge e_{i_2} \wedge \cdots \wedge e_{i_k}$$

$$+ \textstyle\sum_{1 < i_1 < ... < i_k \leq n} c_{i_1 i_2 ... i_k} \; e_{i_1} \wedge e_{i_2} \wedge \cdots \wedge e_{i_k}.$$

Wedging by ξ we get,

$$\xi \wedge \eta = \xi \wedge \sum_{1 < i_1 < ... < i_k \leq n} c_{i_1 i_2 ... i_k} \; e_{i_1} \wedge e_{i_2} \wedge \cdots \wedge e_{i_k}.$$

Now, $\eta \in \mathrm{Ker}(\xi \wedge -)$ implies $\xi \wedge \eta = 0$, which from the equation above implies $c_{i_1 i_2 ... i_k} = 0$, $i_1 > 1$. Therefore

$$\eta = \xi \wedge \left(\sum_{1 < i_2 < ... < i_k \leq n} c_{1 i_2 ... i_k} \; e_{i_2} \wedge \cdots \wedge e_{i_k} \right),$$

which means that $\eta \in \mathrm{Im}(\xi \wedge -)$ and the sequence is exact. $\qquad\square$

From this proposition we see,

Corollary 3.5.16. *The de Rham complex is an elliptic complex.*

Here the cohomology groups are just the de Rham cohomology groups and the index is the usual Euler characteristic, $\chi(M)$.

If $f : M \to M$ is a smooth map, it induces a map

$$f^\star : \Gamma(\textstyle\bigwedge^k (T^\star(M)) \longrightarrow \Gamma(f^\star \textstyle\bigwedge^k (T^\star(M)),$$

given by $\omega(p) \mapsto (p, \omega(f(p)))$, and then the maps φ_k are given by $(p, \omega(f(p))) \mapsto (f^*\omega)(p)$. This lifting then gives induced maps on the cohomology of this complex, denoted $H^k(f^*)$. These are, of course, just the ordinary induced maps of the de Rham cohomology of the manifold M. We know the Lefschetz number of f is given by $L(f) = \sum(-1)^k \operatorname{Tr}(H^k(f^*))$.

Computing $L(f)$ using the Atiyah-Bott formula we get,

$$L(f) = \sum_{p=f(p)} \nu_p$$

where

$$\nu_p = \frac{\sum(-1)^k \operatorname{Tr}(\bigwedge^k (f^*)_p)}{|\det(1 - df_p)|}$$

We now require the following lemma:

Lemma 3.5.17. *For any matrix* $A \in M(n, \mathbb{C})$ *we have*

$$\det(\lambda I - A) = \sum_{k=0}^{n} (-1)^k \lambda^{n-k} \operatorname{Tr}\left(\bigwedge^k A\right).$$

Proof. First suppose A is diagonalizable. Since the characteristic polynomial is independent of the matrix representation, the equation in the statement does not change if we replace A by a conjugate, nor does $\operatorname{Tr}\left(\bigwedge^k A\right)$ change. Hence in this case we can assume that A is diagonal with entries $\lambda_1, \lambda_2, ..., \lambda_n$. If $e_1, ..., e_n$ is the standard basis of $\mathbb{C}^n$, then $Ae_i = \lambda_i e_i$ for any $i = 1, ..., n$ and

$$\det(\lambda I - A) = \chi_A(\lambda) = \prod_{i=1}^{n} (\lambda - \lambda_i).$$

From the right side of the this equation we see the coefficient, c_k, of λ^{n-k} is

$$c_k = (-1)^k \sum_{1 \leq i_1 < \cdots < i_k \leq n} \lambda_{i_1} \cdots \lambda_{i_k}.$$

On the other hand,

$$\bigwedge^k A(e_{i_1} \wedge \cdots \wedge e_{i_k}) = Ae_{i_1} \wedge \cdots \wedge Ae_{i_k} = \lambda_{i_1} e_{i_1} \wedge \cdots \wedge \lambda_{i_k} e_{i_k}$$
$$= \lambda_{i_1} \cdots \lambda_{i_k} (e_{i_1} \wedge \cdots \wedge e_{i_k}),$$

which means that the vectors

$$\left\{ e_{i_1} \wedge e_{i_2} \wedge \cdots \wedge e_{i_k} \mid 1 \leq i_1 < \cdots < i_k \leq n \right\}$$

are eigenvectors of $\bigwedge^k A$ acting on $\bigwedge^k \mathbb{C}^n$. Therefore

$$\mathrm{Tr}\left(\bigwedge^k A \right) = \sum_{1 \leq i_1 < \cdots < i_k \leq n} \lambda_{i_1} \cdots \lambda_{i_k}.$$

Hence, $c_k = (-1)^k \, \mathrm{Tr}\left(\bigwedge^k A \right)$.

Let A be arbitrary. Since both the left and right sides of this equation are polynomials in A, they are each Zariski continuous. Because the diagonalizable matrices are Zariski dense in $M(n, \mathbb{C})$, they agree on A as well.   □

Using this lemma we see that

$$\nu_p = \frac{\det(\mathrm{I} - df_p)}{|\det(\mathrm{I} - df_p)|}.$$

So, finally,

$$L(f) = \sum_{p=f(p)} \frac{\det(\mathrm{I} - df_p)}{|\det(\mathrm{I} - df_p)|} = \sum_{p=f(p)} \mathrm{sgn}\, \det(\mathrm{I} - df_p),$$

and so it coincides with our familiar Lefschetz number (see Definition 3.2.10 and Theorem 3.3.2)[7].

---

[7]A typical application to the complex analytic case is the following: Let X be a connected compact complex manifold with $H^{0;k}(X) = 0$ for $k > 0$. Then any holomorphic map $f : X \to X$ has a fixed point. In particular, if X is complex projective space, any holomorphic automorphism has a fixed point. We shall see connections with this in Chapters 4 and 5.

Chapter 4

Fixed Point Theorems in Geometry

In this chapter we consider both compact and noncompact Riemannian manifolds where we shall look at the relationship between the topology and the curvature. Indeed, each controls the other. In particular, we shall prove the well known fixed point theorem of E. Cartan given just below, as well as certain fixed point theorems in the compact case.

Theorem 4.0.1. Let M a complete, simply connected Riemannian manifold of nonpositive sectional curvature (a Hadamard manifold) and C be a compact group of isometries of M. Then M has a C-fixed point.

Hadamard manifolds include all symmetric spaces G/K, where G is a connected, semisimple Lie group of noncompact type and K is a maximal compact subgroup of G. Theorem 4.0.1 is proved in this important special case in [1], p. 293, and in [64], p. 111 (volume 2) in general. Of course, Hademard manifolds also include $\mathbb{R}^n$ (and C can be any compact group of isometries). As we saw in Corollary 1.4.4 these act affinely, so here we recapture Theorem 1.4.1. All known proofs of Theorem 4.0.1, or any of its special cases require the concept of *center of gravity* by averaging via Haar measure over the compact group C.

4.1   Some Generalities on Riemannian Manifolds

For the reader's convenience we now review some notions from Riemannian Geometry which will be needed in the sequel.

In Chapter 3 we considered a manifold M is and its tangent bundle $T(M)$. A section $X \in \Gamma(T(M))$ is a *vector field*. Now, X defines a first order differential operator ∇_X by $\nabla_X f(p) = \nabla_{X(p)} f(p)$. One sees easily that ∇_X is a derivation (i.e. linear over $C^\infty(M)$, that is $\nabla_{gX+Y}\, f = g\nabla_X\, f + \nabla_Y\, f$ for any f, $g \in C^\infty(M)$, and satisfies the Leibnitz rule $\nabla_X(fg) = f\nabla_X\, g + (\nabla_X\, f)g$.

We note the correspondence $Xf \mapsto \nabla_X f$ is a $C^\infty(M)$-linear operator on $\Gamma(T(M))$. This operator can be identified with the section $df \in \Gamma(T^\star(M))$ of the cotangent bundle $T^\star(M)$. Therefore, we write

$$\nabla_X\, f = df(X).$$

Although there is a natural way to differentiate a smooth function $f : M \to \mathbb{R}$ with respect to a tangent vector, there is no natural way to differentiate vector fields. Actually, a fixed rule for the differentiation of vector fields is itself an additional structure to be placed on M called an *affine connection*.

Definition 4.1.1. An affine connection (or simply a connection) on M is a mapping which assigns to any two smooth vector fields X and Y a third one $\nabla_X\, Y$ called the *covariant derivative* of the vector field Y with respect to the vector field X, having the following properties:

1. ∇ is $\mathbb{R}$-linear in Y and C^∞-linear in X.

2. $\nabla_{fX+Y}\, Z = f\nabla_X\, Z + \nabla_Y\, Z$.

3. $\nabla_X fY = (Xf)Y + f\nabla_X Y$.

If $(U, (x_1, \ldots, x_n))$ is a chart on M, ∇ a connection on M, and $\partial_i := \frac{\partial}{\partial x_i}$, $i = 1, ..., n$ are the corresponding basic vector fields on U, then

$$\nabla_{\partial_i}\, \partial_j = \sum_{k=1}^{n} \Gamma_{ji}^k \partial_k,$$

where the coefficients Γ^k_{ji} are smooth functions called *Christoffel symbols*.

Let $\gamma : [0,1] \to M$ be a smooth curve in M. Henceforth we will use equivalently and without distinction, the notations $\dot{\gamma}$, γ', $\frac{d}{dt}\gamma$, or $\frac{\partial}{\partial t}\gamma$ for the derivative of $\gamma(t)$.

Definition 4.1.2. A vector field X *along the curve* γ is a function which assigns to each $t \in \mathbb{R}$ a tangent vector $X_t \in T_{\gamma(t)}M$, such that for any smooth function f on M the correspondence $t \mapsto X_t f$ is a smooth function on $\mathbb{R}$.

For example the velocity vector $\frac{d\gamma}{dt}$ is the vector field along γ defined by

$$\frac{d\gamma}{dt} := \gamma_\star\left(\frac{d}{dt}\right), \quad \text{where } \gamma_\star : T_t\mathbb{R} \to T_{\gamma(t)}M.$$

Suppose M has an affine connection. Then any vector field X along a curve γ determines a new vector field $\frac{DX}{dt}$ also along γ, called the covariant derivative of X.

Definition 4.1.3. A vector field X along the curve γ is said to be a *parallel vector field* if the covariant derivative $\frac{D}{dt}X \equiv 0$.

Definition 4.1.4. A connection is called *symmetric* or *torsion free* if

$$\nabla_X Y - \nabla_Y X = [X, Y].$$

Applying this relation when $X = \partial_i$ and $Y = \partial_j$, since $[\partial_i, \partial_j] = 0$, we get

$$\Gamma^k_{ij} = \Gamma^k_{ji}.$$

We remark that the converse is also true, however we shall not need this.

Definition 4.1.5. If M is manifold and for each $p \in M$ the tangent space $T_p(M)$ is equipped with a positive definite symmetric bilinear form $\langle,\rangle_p$, so that for any two smooth vector fields X, Y the function $M \to \mathbb{R}$ defined by $p \mapsto \langle X_p, Y_p \rangle_p$ is smooth, then we shall say M is a *Riemannian manifold* and we call the set g_p of bilinear forms on the various tangent spaces the *Riemannian metric* of M.

Definition 4.1.6. We say that the connection ∇ on M is *compatible with the Riemannian metric g* if parallel translation preserves inner products. In other words, for any curve γ and any pair X, Y of parallel vector fields along γ, the inner product $\langle X, Y \rangle$ is constant (i.e. $\nabla g = 0$).

Now, if M is a Riemannian manifold, we can introduce differentiation of vector fields in a natural way. Indeed, we have

Theorem 4.1.7. (Fundamental Theorem of Riemannian Geometry). *A Riemannian manifold has one and only one symmetric connection which is compatible with its metric.*

This important fact is proved in [70], p. 48. The unique symmetric affine connection which is compatible with the metric on M is called the *Levi-Civita connection*.

Let M be a Riemannian manifold and $g := \langle \cdot \, , \, \cdot \rangle$ be the metric tensor. Let $\Gamma(T(M))$ denote the space of sections of $T(M)$, and ∇ denote the Levi-Civita connection of M.

Definition 4.1.8. For any three vector fields X, Y, Z we define the *Riemannian curvature R on M* as a section:

$$R \in \Gamma \left(\operatorname{Hom} \left(\bigwedge^2 T(M), \operatorname{Hom}\left(T(M), T(M)\right) \right) \right)$$

defined by

$$
\begin{aligned}
(X, Y, Z) \to R(X, Y, Z) =&: R(X, Y)Z \\
=& \nabla_X \nabla_Y Z - \nabla_Y \nabla_X Z - \nabla_{[X,Y]} Z.
\end{aligned}
$$

One sees easily that R is indeed a section of

$$\operatorname{Hom} \left(\bigwedge^2 T(M), \operatorname{Hom}\left(T(M), T(M)\right) \right)$$

and (see [70], p. 53) that the Riemannian curvature R satisfies:

1. $R(X, Y)Z = -R(Y, X)Z$.

2. $R(X,Y)Z + R(Y,Z)X + R(Z,X)Y = 0$ (the *Bianchi identity*).

3. $\langle R(X,Y)Z, W \rangle + \langle R(X,Y)W, Z \rangle = 0$.

4. $\langle R(X,Y)Z, W \rangle = \langle R(Z,W)X, Y \rangle$.

From the Riemannian curvature tensor R we obtain several notions of curvature[1]:

1. **The sectional curvature** $K(\cdot, \cdot)$. For every linearly independent pair of vectors X, $Y \in T_p(M)$, we define

$$K(X,Y) = \frac{\langle R(X,Y)Y, X \rangle}{\|X\|^2 \, \|Y\|^2 - \langle X, Y \rangle^2}.$$

K is defined on the Grassmann space of all two dimensional linear subspaces of $T_p(M)$ (depending only on the linear span of (X,Y)).

2. The **Ricci curvature**[2], $\mathrm{Ric}(X,Y)$, is defined by

$$\mathrm{Ric}(X,Y) := \mathrm{Tr}\left(Z \to R(Z,X)Y\right).$$

Using the curvature identities above we see that

$$\mathrm{Ric}(X,Y) = \mathrm{Ric}(Y,X).$$

---

[1]See Tao, [95], p. 372. The relationship between the Riemannian, Ricci, and scalar curvatures depends on the dimension as follows:

1. In dimension one, all three curvatures vanish.

2. In two dimensions, the Riemannian and Ricci curvatures are each just multiples of the scalar curvature (by some tensor depending algebraically on the metric).

3. In three dimensions, the Riemann tensor is a linear combination of the Ricci curvatures in the three fundamental directions.

4. In four and higher dimensions, the Riemann tensor is not fully controlled by the Ricci curvature; there is an additional component to the Riemann tensor, namely the Weyl tensor. Similarly, the Ricci curvature is not fully controlled by the scalar curvature.

[2]Although we do not need the notions of Ricci and scalar curvatures to prove E. Cartan's fixed point theorem, it seems to the authors that having these in compact and accessible form might be useful.

The Ricci curvature in the direction X (where X is a unit vector) is given by

$$\mathrm{Ric}(X) = \mathrm{Ric}(X, X).$$

3. The **scalar curvature** s is defined as

$$s = \mathrm{Tr}(\mathrm{Ric}) = \sum \mathrm{Ric}(E_i, E_i),$$

where E_i, $i = 1, ..., n$, is any local orthonormal basis.

There is a close relationship between $R_Z := R(\cdot, Z)Z$ and the sectional curvature: Indeed, take Z with $\|Z\| = 1$. For X orthogonal to Z

$$\langle R_Z X, X \rangle = \langle R(X, Z)Z, X \rangle = K(Z, X)\|X\|^2.$$

Geometrically, positive scalar curvature means that infinitesimal balls have slightly smaller volume than in the Euclidean case; positive Ricci curvature means that infinitesimal sectors have slightly smaller volume than in the Euclidean case; and positive sectional curvature means that all infinitesimally geodesic two-dimensional surfaces have positive Gaussian curvature.

In the nonpositive case the relationship between these various curvatures in any dimension is as follows: Nonpositive Riemannian curvature $\Longrightarrow$ nonpositive sectional curvature $\Longrightarrow$ nonpositive Ricci curvature $\Longrightarrow$ nonpositive scalar curvature[3].

Because there are all these different notions of curvature, we now need to fix on the one we will be using when we say (M, g) has nonpositive curvature at a point p. This will always be *sectional curvature*.

In order to prove E. Cartan's fixed point theorem we shall need to understand how the geometry of a Riemannian manifold is controlled locally by the curvature tensor and for that we introduce the notion

---

[3]See [95], p. 373. In dimension one, this is vacuously true. In two dimensions; these conditions are all equivalent; and in three dimensions, nonpositive Riemannian curvature is equivalent to nonpositive sectional curvature (because every 2-form is the wedge product of two one-forms). But these conditions are otherwise inequivalent. In four and higher dimensions each of these conditions is distinct. Analogous statements hold in the case of nonnegative curvature.

of a Jacobi field. As we shall see, this is a particularly useful tool in understanding the dependence of local distance on the curvature. For this we require,

Definition 4.1.9. A smooth 1-*parameter family of geodesics* in the Riemannian manifold M is a smooth map $F : (-\epsilon, \epsilon) \times [0, 1] \to M$ such that the curve $\gamma_t : [0, 1] \to M$ given by $\gamma_t : s \mapsto F(t, s)$ is a geodesic for all $t \in (-\epsilon, \epsilon)$.

Now, let $F(t, s)$ be a 1-parameter family of geodesics in M and for each $t \in (-\epsilon, \epsilon)$ consider the vector field J_t along γ_t given by

$$J_t(s) = \frac{\partial F}{\partial t}(t, s).$$

Then,

Proposition 4.1.10. *The vector field J_t satisfies the second order differential equation*

$$\nabla_{\gamma_t'} \nabla_{\gamma_t'} J_t + R(J_t, \gamma_t')\gamma_t' = 0,$$

where R is the Riemannian curvature tensor of M.

Proof. Set $J(t, s) = J_t(s) = \frac{\partial F}{\partial t}$ and $X(t, s) = \frac{\partial F}{\partial s}$. Since $[\frac{\partial}{\partial t}, \frac{\partial}{\partial s}] = 0$ we get

$$[J, X] = \left[dF\left(\frac{\partial}{\partial t}\right), dF\left(\frac{\partial}{\partial s}\right) \right] = dF\left(\left[\frac{\partial}{\partial t}, \frac{\partial}{\partial s}\right] \right) = 0.$$

In addition $\nabla_X X = 0$, since F is a family of geodesics. Therefore, by the definition of the Riemannian curvature R, we get

$$R(J, X)X = \nabla_J \nabla_X X - \nabla_X \nabla_J X - \nabla_{[J,X]} X$$
$$= -\nabla_X \nabla_J X = -\nabla_X \nabla_X J.$$

Therefore,

$$\nabla_X \nabla_X J + R(J, X)X = 0$$

and this holds for all t.   □

Now, for a geodesic $\gamma : [0,1] \to M$, setting $X = \frac{d\gamma}{dt}$, we define

Definition 4.1.11. A vector field J along a geodesic γ is called a *Jacobi field* if it satisfies the Jacobi differential equation

$$\frac{D^2 J}{dt} + R\left(\frac{d\gamma}{dt}, J\right)\frac{d\gamma}{dt} = 0.$$

Let $\gamma : [a,b] \to M$ be a geodesic starting from the point $p = \gamma(a)$. Then,

Definition 4.1.12. The point $q = \gamma(b)$ in M is called *conjugate point* of p along γ if there exists a nonzero Jacobi field $J(t)$ along γ with $J(a) = 0 = J(b)$.

We remark that in dimension n, there exist exactly n linearly independent Jacobi fields along γ, which are not zero at $p = \gamma(a)$. This follows from the fact that the Jacobi fields $J_1, ..., J_k$ with $J_i(a) = 0$ are linearly independent if and only if $J_1'(a), ..., J_k'(a)$ are linearly independent. In addition, the Jacobi field $J(t) = t\gamma'(t) \neq 0$ for any $t \neq a$. From this we conclude that the multiplicity of a conjugate point is always $\leq n - 1$.

Roughly speaking, conjugate points are those which can almost be joined by 1-parameter families of geodesics. For example, on a sphere, antipodal points can be joined by an infinity of great circles so they are conjugate. On $\mathbb{R}^n$, there are no conjugate points. More generally, on Riemannian manifolds with nonpositive sectional curvature, there are no conjugate points. Indeed, we have the following proposition which will play an important role in the proof of Cartan's theorem:

Proposition 4.1.13. *If the sectional curvature of a Riemannian manifold M is nonpositive at every point, then no two points of M are conjugate along a geodesic.*

Proof. Let M be a Riemannian manifold with sectional curvature $K \leq 0$ at any point, and $p = \gamma(a)$ be the initial point of the geodesic $\gamma : [a,b] \to M$, *parametrized by arclength*. Moreover, suppose that J is a Jacobi field along γ with $J(t) \neq 0$ for all $t \in (a,b)$ and $J(a) = 0 = J(b)$. Then,

$$\langle J(t), \gamma'(t) \rangle = 0,$$

for any $t \in [a, b]$. Now, since $K \leq 0$

$$\frac{d}{dt}\left\langle \frac{DJ}{dt}, J \right\rangle = \left\langle \frac{D^2 J}{dt^2}, J \right\rangle + \left\langle \frac{DJ}{dt}, \frac{DJ}{dt} \right\rangle = -K\left\langle J, J \right\rangle + \left| \frac{DJ}{dt} \right|^2 \geq 0,$$

which implies that $\langle \frac{DJ}{dt}, J \rangle$ is an increasing function. Since by hypothesis $J(a) = J(b) = 0$ it follows that for $t = a$ and $t = b$, $\langle \frac{DJ}{dt}, J \rangle = 0$. The mean value theorem then tells us $\langle \frac{DJ}{dt}, J \rangle = 0$ on $[a, b]$. Eventually, since $\frac{d}{dt}\langle J, J \rangle = 2\langle \frac{DJ}{dt}, J \rangle = 0$, we get $\langle J, J \rangle = $ constant, Taking into consideration that $J(a) = 0$, we obtain

$$|J(t)| = 0 \quad \text{for all } t \in [a, b],$$

a contradiction. Therefore M has no conjugate points. □

We now define the exponential map of M at p. This allows us to compare neighborhoods of p in M with those about 0 in $T_p(M)$. We note that this is not the same thing as the exponential map of a Lie group.

Definition 4.1.14. If p is any point of a manifold M, we define the *exponential map* $\exp_p$ as

$$\exp_p : T_p(M) \to M, \quad \xi \mapsto \exp_p(\xi) := \gamma_\xi(1)$$

for all $\xi \in T_p(M)$ such that γ_ξ is defined on the interval $[0, 1]$ and it is the geodesic of M whose initial velocity is ξ.

It is customary to call a subset S of a real vector space V *star shaped* about 0 if $v \in S$ implies $tv \in S$, where $t \in [0, 1]$. In other words, S is the union of all radial line segments joining 0 and its other points. For a manifold M and tangent space $T_p(M)$,

Definition 4.1.15. If $\widetilde{U}$ is a star shaped neighborhood of 0 in $T_p(M)$ and $\exp_p$ is a diffeomorphism of $\widetilde{U}$ onto a neighborhood $\widetilde{U}$ of p, then, $\widetilde{U}$ is called a *normal neighborhood* of p.

Remark 4.1.16. Consider a point $p \in M$ and assume that $\exp_p$ is defined at $v \in T_p(M)$ and take $w \in T_v(T_p(M))$. Consider the curve $v(s)$ in $T_p(M)$, where $\epsilon > 0$ and $s \in (-\epsilon, \epsilon)$ starting at v with initial velocity w (i.e. $v(0) = v$ and $\frac{d}{ds}v(s)|_{s=0} = w$). Now the surface

$$F : [0,1] \times [-\epsilon, \epsilon] \to M, \quad (t,s) \mapsto F(t,s) := \exp_p tv(s)$$

has second derivative $\frac{\partial}{\partial s}F(t,0) = (\exp_p)_*(tv)(tw)$, and setting

$$J(t) = \frac{\partial}{\partial s}F(t,s)|_{s=0},$$

as we saw just above, $J(t)$ is a Jacobi field along the geodesic $\gamma(t) = \exp_p(tv)$, $t \in [0,1]$, with $J(0) = 0$.

We remark this is actually the only way of constructing Jacobi fields along geodesics, with $J(0) = 0$ and $J'(0) = \dot{v}(0)$ (see do Carmo, [32]).

The following three lemmas, which will be needed later, show the importance of the exponential map.

Lemma 4.1.17. *The exponential map $\exp_p$ carries radial lines from $\vec{0} \in T_p(M)$ to geodesics starting at $p \in M$. More explicitly,*

$$\exp_p(t\xi) = \gamma_\xi(t)$$

for each t for which $\exp_p(t\xi)$ is well defined.

Proof. Fix a $t \in \mathbb{R}$ and a tangent vector $\xi \in T_p(M)$. The geodesic $s \to \gamma_\xi(ts)$ has initial velocity $t\frac{d\gamma_\xi}{ds}(0)$. But this initial velocity has the geodesic $\gamma_{t\xi}$. Therefore, by the local uniqueness of solutions of differential equations with an initial condition, these two geodesics are equal, i.e.

$$\gamma_\xi(ts) = \gamma_{t\xi}(s)$$

whenever both sides are well defined. Setting $s = 1$ we get

$$\gamma_\xi(t) = \gamma_{t\xi}(1) = \exp_p(t\xi).$$

$\square$

The next lemma tells us $\exp_p$ is a local diffeomorphism. The neighborhood we get is called a *normal neighborhood* of p in M.

Lemma 4.1.18. *For each $p \in M$, $\exp_p$ sends some neighborhood U of $\vec{0} \in T_p(M)$ diffeomorphically onto a neighborhood N of p.*

Proof. It suffices to show that the differential of $\exp_p$ at $\vec{0}$ is a linear isomorphism $(\exp_p)_\star : T_p(M) \to T_p(M)$ (here we identify $T_0(T_p(M))$ with $T_p(M)$). This will imply $\exp_p$ is a diffeomorphism of some neighborhood of $\vec{0}$, by the inverse function theorem.

A tangent vector ξ to $T_p(M)$ at $\vec{0}$ is the initial velocity of the ray $\sigma(t) = t\xi$, and as proved in Lemma 4.1.17

$$\exp_p(\sigma(t)) = \gamma_\xi(t).$$

Since tangent maps preserve velocities

$$(\exp_p)_\star(\xi_0) = (\exp_p)_\star(\sigma'(0)) = \gamma_\xi'(0) = \xi.$$

This means the tangent map of $\exp_p$ at $\vec{0}$ is just the natural isomorphism $\xi_0 \to \xi$. $\qquad\square$

Lemma 4.1.19. *A point $p \in M$ has a conjugate in M if and only if $(\exp_p)_\star$ is singular.*

Proof. Let $p \in M$, $\gamma : [0, a] \to M$ a geodesic in M with $\gamma(0) = p$ and consider a point $q = \gamma(t_0)$. The point q is conjugate to p along the geodesic γ if and only if there is a nonzero Jacobi field J along γ such that $J(0) = J(t_0) = 0$. Now, set $v = \dot{\gamma}(0)$ and $w = \dot{v}(0)$, and consider (by Remark 4.1.16), the Jacobi field along γ, given by

$$J(t) = (\exp_p)_\star(tv)(tw), \quad t \in [0, a].$$

Then $J(t) \neq 0$ if and only if $w \neq 0$. Hence, q is conjugate to p if and only if

$$0 = J(t_0) = (\exp_p)_\star(t_0 v)(t_0 w)$$

with $w \neq 0$, in other words if and only if $t_0 v$ is a critical point of $\exp_p$. $\qquad\square$

When the manifold M has nonpositive curvature an important consequence of the last lemma, together with Proposition 4.1.13 telling us M has no conjugate points is:

Corollary 4.1.20. *If M has nonpositive curvature, $(\exp_p)_\star$ is nonsingular at every point p.*

Definition 4.1.21. The *injectivity radius* at a point p of a Riemannian manifold is the largest radius ball on which the exponential map at p is a local diffeomorphism.

In other words, the injectivity radius for a point p is the largest distance m such that any geodesic starting from p is length-minimizing for at least a distance m. Thus, if M is a complete manifold, and the injectivity radius at p is a finite number m, then either there is a geodesic of length $2m$ which starts and ends at p or there is a point q conjugate to p at a distance m from p. If M is a compact Riemannian manifold, then the injectivity radius is either half the minimal length of a closed geodesic, or the minimal distance between conjugate points on a geodesic.

We now define the length metric on M.

Definition 4.1.22. Given a path $\gamma : [0,1] \to M$, we define its *length* as

$$L(\gamma) := \int_0^1 |\gamma^{'}(t)| dt.$$

Also, we define the *distance function d* on M by

$$d(p,q) := \inf_\gamma L(\gamma),$$

where the infimum is taken over all piecewise smooth paths $\gamma : [0,1] \to M$, with $\gamma(0) = p$ and $\gamma(1) = q$. A geodesic segment is called *minimizing* if its length is equal to the distance between the two endpoints.

Remark 4.1.23. In general, in a Riemannian manifold M, having the two points p and q, $d(p,p) = l$ does not always mean that there is a curve connecting them of length l. One sees this easily in the Riemannian manifold $\mathbb{R}^2 - \{0\}$ with points $p = (0, 1/2)$ and $q = (0, -1/2)$. Later we will see that, in a normal neighborhood, the distance to the center point is always attained by radial geodesics. In general, if the distance between points is attained by a curve, then this curve must be a geodesic (see [32], p. 70).

We mention that equipped with this metric, (M, d) becomes a metric space. A proof of this can be found in [62], p. 14.

Definition 4.1.24. The *energy* $E(\gamma)$ of the curve $\gamma : [0, 1] \to M$ is defined as

$$E(\gamma) := \int_0^1 |\dot{\gamma}(t)|^2 dt.$$

Remark 4.1.25. The advantage in dealing with $E(\gamma)$ is that the energy functional is smooth (since it lacks the square root), while the length functional $L(\gamma)$ is smooth only at those curves $\gamma(t)$ that satisfy $\frac{d}{dt}\gamma(t) \neq 0$ for all t.

Definition 4.1.26. We say that a Riemannian manifold M is *(geodesically) complete* if all geodesics are defined for all t; i.e. if $\exp_p(tX)$ is defined for all $X \in T_p(M)$, $t \in \mathbb{R}$ and for all $p \in M$.

It then follows from the preceding proposition that if M is geodesically complete, then any two of its points can be joined by a minimizing geodesic.

The following result is of fundamental importance and, of course, we shall make frequent use of it in the sequel. It states that the different concepts of completeness are equivalent for a Riemannian manifold M.

Theorem 4.1.27. (Hopf-Rinow Theorem). *The following are equivalent:*

1. *M is complete as a metric space under d, i.e. every Cauchy sequence converges.*

2. *There exists a point $p \in M$ such that $\exp_p$ is defined on the entire $T_p(M)$.*

3. M is geodesically complete.

4. Every closed d-bounded subset of M is compact.

Proof. That 3) implies 2) is trivial. Also, 4) implies 1) since a Cauchy sequence is bounded. Hence if 4) holds, its closure is compact, so the Cauchy sequence contains a convergent subsequence and hence the original sequence converges. Therefore, we must prove that 1) implies 3) and 2) implies 4).

1) $\Rightarrow$ 3). Let $\gamma : [0, b) \to M$ be a unit speed geodesic. We must show that γ can be extended past b. Let $t_i \to b$. Then $d(\gamma(t_i), \gamma(t_j)) \le |t_i - t_j|$, so $\gamma(t_i)$ forms a Cauchy sequence and hence, by completeness under d, converges to some point $q \in M$. Then $d(\gamma(s), q) \le d(\gamma(s), \gamma(t_i)) + d(\gamma(t_i), q)$ so the curve $\gamma(s)$ approaches q as $s \to b$. Let C be a convex neighborhood of q. For some t close enough to b the point $w := \gamma(t)$ lies in C and hence γ restricted to $[t; b)$ is a radial geodesic emanating from w. But it can then be extended as a radial geodesic until it hits the boundary of C, which means that γ can be extended as a geodesic past b.

2) $\Rightarrow$ 4). Let p be a point for which $\exp_p$ is defined on the entire T_pM. 2) tells us that this means $\exp_q$ is geodesically complete at every point of $q \in M$. Hence there is a minimizing geodesic $\gamma_q : [0, 1] \to M$ joining p to q. If q ranges over a bounded closed subset A of M, the set $d(p, q) = \| \frac{d}{dt}\gamma_q|_{t=0} \|$ is bounded, and hence the set of such $\frac{d}{dt}\gamma_q|_{t=0}$ lies in some closed ball B in $T_p(M)$, which is compact. Hence $\exp_p(B)$ is compact. But $A \subseteq \exp_p(B)$ and a closed subset of a compact set is compact. □

Corollary 4.1.28. *Any compact manifold M is geodesically complete. Therefore at any point $p \in M$, the exponential map is defined on the whole tangent space $T_p(M)$.*

Corollary 4.1.29. *If M is a complete Riemannian manifold then any two points p and q can be connected by a minimizing geodesic.*

Proof. Since M is complete, by the Hopf-Rinow Theorem, $\exp_p$ is defined on all $T_p(M)$, for any $p \in M$. Let q be another point in M and

$l = d(p, q)$. Let $B_\epsilon(p)$ be a normal ball in M (i.e. the image under $\exp_p$ of a ball $\widetilde{B}_\epsilon$ at the origin in $T_p(M)$) and set $S_\epsilon(p) \equiv \partial B_\epsilon(p)$ (the boundary of $B_\epsilon(p)$). Clearly $S_\epsilon(p)$ is compact since it is bounded and closed. Therefore, the continuous map $f(x) := d(q, x)$, $x \in S_\epsilon(p)$, achieves a minimum at some point $x_0 \in S_\epsilon(p)$. Let $x_0 = exp_p(\epsilon v)$ where $v \in T_p(M)$ with $|v| = 1$. Consider the geodesic $\gamma(t) = exp_p(tv)$, $t \in [0, \infty)$. We claim $\gamma(l) = q$. To prove this, consider the equation

$$d(\gamma(t), q) = l - t. \tag{4.1}$$

Let $\mathcal{R}$ be the set of solutions of the equation (4.1), for $0 \le t \le l$. Obviously, $\mathcal{R} \ne \emptyset$ since $t = 0$ is in $\mathcal{R}$. We want to prove that $l \in \mathcal{R}$. To do this, first we observe that $\mathcal{R}$ is a closed set in $[0, l]$. Take any $t_0 \in \mathcal{R}$. We want to prove that if $t_0 < l$, then for some sufficiently small $\delta > 0$, $t_0 + \delta$ is also a solution of equation (4.1). This will imply that $\sup \mathcal{R} = l$, and since $\mathcal{R}$ closed, $l \in \mathcal{R}$, i.e. $\gamma(l) = q$.

To prove $(t_0 + \delta) \in \mathcal{R}$, again let $B_\delta(\gamma(t_0))$ be a normal ball at $\gamma(t_0)$ and $S_\delta \equiv \partial B_\delta(\gamma(t_0))$. Denote by $\bar{x}_0$ the point of S_δ where the function $d(x, q)$, achieves a minimum. It suffices to show that

$$\bar{x}_0 = \gamma(t_0 + \delta).$$

Indeed, since

$$d(\gamma(t_0), q) = \delta + \min d(x, q) = \delta + d(\bar{x}_0, q),$$

and

$$d(\gamma(t_0), q) = l - t_0,$$

if $\bar{x}_0 = \gamma(t_0 + \delta)$, we get

$$l - t_0 = \delta + d(\bar{x}_0, q) = \delta + d(\gamma(t_0 + \delta), q), \tag{4.2}$$

which implies

$$d(\gamma(t_0 + \delta), q) = l - (t_0 + \delta).$$

In other words $(t_0 + \delta) \in \mathcal{R}$.

We prove $\bar{x}_0 = \gamma(t_0 + \delta)$. From the triangle inequality

$$d(p, q) \leq d(p, \bar{x}_0) + d(\bar{x}_0, q),$$

and the left side of equation (4.2) we get

$$d(p, \bar{x}_0) \geq d(p, q) - d(\bar{x}_0, q) = l - (l - t_0 - \delta) = t_0 + \delta.$$

On the other hand, the broken curve joining p to $\bar{x}_0$ goes from p to $\gamma(t_0)$ via the geodesic γ and from $\gamma(t_0)$ to $\bar{x}_0$ by the geodesic ray, has length $t_0 + \delta$.

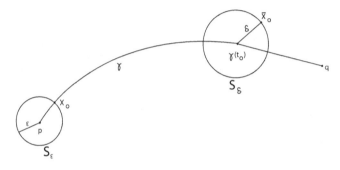

Therefore,

$$d(p, \bar{x}_0) = t_0 + \delta,$$

and such a curve must be a distance minimizing geodesic since it has length less or equal to the length of any other curve joining these to points. $\qquad \square$

4.2   Hadamard Manifolds and Cartan's Theorem

We now come to the so called Hadamard theorem[4].

---

[4]According to Shlomo Sternberg this theorem was first proved by von-Mangoldt in 1881 (J. Reine Angew. Math. 91) for surfaces. Then Hadamard gave two more proofs for surfaces in 1889 (see J. Math. Pur. et App.), with some hint on how to do it in three dimensions. Cartan proved the full theorem in n-dimensions in 1922. So, naturally it is called Hadamard's theorem since he was neither the first to prove it for surfaces, nor did he prove it in full generality.

Here, the condition on the sectional curvature in the hypothesis is crucial, since it is the sectional curvature which governs the behavior of geodesics. In fact, we shall see that positive sectional curvatures cause geodesics emanating from a point $p \in M$ to converge, while nonpositive sectional curvatures cause them to diverge and it is this fact which underlies the proof of what perhaps should be called the Hadamard-Cartan Theorem.

Now, let M, and $\widetilde{M}$ be two Riemannian manifolds.

Definition 4.2.1. We say that the smooth map $\pi : \widetilde{M} \to M$ is a smooth *covering map* when,

1. π is onto, and

2. for each $p \in M$ there is an open neighborhood U_p such that $\pi^{-1}(U_p)$ is a disjoint union of open sets in $\widetilde{M}$ each of which is diffeomorphic to U_p via π.

Lemma 4.2.2. *Let M and $\widetilde{M}$ be two Riemannian manifolds with $\widetilde{M}$ complete. Then, any local isometry $\pi : \widetilde{M} \to M$ is a smooth covering map.*

Proof. Let $p \in M$. We have to find an open neighborhood U_p of p such that $\pi^{-1}(U_p)$ is the disjoint union of (open) diffeomorphic copies of U_p. To do this we proceed as follows. By Lemma 4.1.18, we can find $\epsilon > 0$ such that $\exp_p$ sends the open ball $B(p, 2\epsilon)$ in $T_p(M)$ diffeomorphically onto the open subset, $\{q \in M \mid d(q, p) < 2\epsilon\}$ in M. Consider also the fiber

$$\pi^{-1}(\{p\}) = \{\widetilde{p}_i \in \widetilde{M}, \ i \in I \mid \pi(\widetilde{p}_i) = p\}.$$

Now, set

$$U_p := \{q \in M \mid d(q, p) < \epsilon\}, \quad \widetilde{U}_i := \{\widetilde{q} \in \widetilde{M} \mid d(\widetilde{q}, \widetilde{q}_i) < \epsilon\}.$$

Pick a point $\widetilde{p}_k$ in the fiber over p and consider the two open balls $B(\epsilon) = \{X \in T_p(M) \mid \|X\| < \epsilon\}$ and $\widetilde{B}(\epsilon) = \{Y \in T_{\widetilde{p}_k}(\widetilde{M}) \mid \|Y\| < \epsilon\}$, in the tangent spaces $T_p(M)$ and $T_{\widetilde{p}_k}(\widetilde{M})$ respectively. Clearly the following diagram:

$$\widetilde{B}(\epsilon) \xrightarrow{\ \pi_* \ } B(\epsilon)$$

$$exp_* \Big\downarrow \qquad\qquad \Big\downarrow exp$$

$$\widetilde{U}_k \xrightarrow{\ \ \pi \ \ } U_p$$

is commutative, (here the map $exp_* : \widetilde{B}(\epsilon) \to \widetilde{U}_k$ is defined on all $T_{\widetilde{p}_k}(\widetilde{M})$ because by assumption $\widetilde{M}$ is complete). The open set U_p is the neighborhood we are looking for. To see this we have to check, first that $\widetilde{U}_{\widetilde{p}_k} \cap \widetilde{U}_{\widetilde{p}_j} = \emptyset$ if $i \neq j$. Indeed, let $\widetilde{U}_{\widetilde{p}_k} \cap \widetilde{U}_{\widetilde{p}_j} \neq \emptyset$, and take a point $\widetilde{s} \in \widetilde{U}_{\widetilde{p}_k} \cap \widetilde{U}_{\widetilde{p}_j} \neq \emptyset$. Let $\widetilde{\gamma}_i$ be the geodesic joining $\widetilde{p}_i$ with $\widetilde{s}$, and $\widetilde{\gamma}_j$ the geodesic between $\widetilde{p}_j$ and $\widetilde{s}$ (both exist because of the chosen neighborhood of radius $< \epsilon$). Now, both are projected onto the geodesic $\pi(\widetilde{\gamma}_i)$ and $\pi(\widetilde{\gamma}_j)$, both of them connect $\pi(\widetilde{s})$ with p, and both have length $< \epsilon$. Since in this neighborhood, the geodesic joining two points is unique, we must have $\pi(\widetilde{\gamma}_j) = \pi(\widetilde{\gamma}_i)$. Since by assumption π is a local isometry, $\widetilde{\gamma}_i = \widetilde{\gamma}_j$ which implies $\widetilde{p}_i = \widetilde{p}_j$, i.e. $i = j$.

Now, take a point $\widetilde{s} \in \pi^{-1}(U)$ and set $s = \pi(\widetilde{s}) \in U$. We know that there is geodesic γ joining s to p in U, of unit speed and of length $< \epsilon$. Since π is a local isometry, γ can be lifted to a unit-speed geodesic $\widetilde{\gamma}$ in $\widetilde{M}$ starting at $\widetilde{s}$ so that $\pi \circ \widetilde{\gamma} = \gamma$ and joining $\widetilde{s}$ to some $\widetilde{p}_k$ with length $< \epsilon$. Hence $\widetilde{s} \in \widetilde{U}_k$, for some k. All this means that

$$\pi^{-1}(U) = \bigcup_{i \in I} \widetilde{U}_i.$$

From the above we see $\pi(M)$ is open and closed in M. Since M and $\widetilde{M}$ are connected π is surjective. $\qquad\qquad\qquad\qquad\qquad\qquad \square$

We are now ready to state and prove Hadamard's Theorem. It explains why a symmetric space of noncompact type can always be modeled on an open disk.

Theorem 4.2.3. (Hadamard's Theorem.) *Let M be a complete Riemannian manifold with everywhere nonpositive sectional curvature. Then for every $p \in M$ the exponential map $\exp_p : T_p(M) \to M$ is a* *smooth covering map. If in addition M is simply connected, $\exp_p$ is a diffeomorphism. Thus M is diffeomorphic to Euclidean space[5].*

Proof. Because the sectional curvature of M is nonpositive at every point, by Proposition 4.1.13, there are no conjugate points along any geodesic. Since $(\exp_p)_\star$ is nonsingular at every point $v \in T_p(M)$, we can define a Riemannian metric $\widetilde{g}$ on $T_p(M)$ by

$$\widetilde{g}(x,y) := g((\exp_p)_\star(x), (\exp_p)_\star(y)), \quad \text{for all } x,\, y \,\in\, T_v(T_p(M)).$$

Now $\exp_p$ sends lines through the origin in $T_p(M)$ to geodesics through $p \in M$. Hence, straight lines through the origin must be geodesics in the Riemannian manifold $(T_p(M), \widetilde{g})$. From the Hopf-Rinow theorem it follows that $(T_p(M), \widetilde{g})$ is complete. Hence, by Lemma 4.2.2, $\exp_p$ is a covering map. Moreover, by the Homotopy Lifting Theorem, if M is simply connected, $\exp_p$ is a diffeomorphism. $\square$

This result has the following important consequence.

Corollary 4.2.4. *If M is complete and has sectional curvature ≤ 0, the universal cover $\widetilde{M}$ is diffeomorphic to $\mathbb{R}^n$. In particular, if M is complete, simply connected and has sectional curvature ≤ 0, M cannot be compact.*

We now formalize things.

Definition 4.2.5. A Riemannian manifold which is simply connected, complete, and has nonpositive sectional curvature is called a *Hadamard manifold*[6].

---

[5]In contrast, a very recent result of Böhm and Wilking ([13]) shows a compact, simply connected Riemannian manifold of positive curvature is diffeomorphic to a sphere.

[6]From the topological viewpoint a Hadamard manifold is a very simple object. Actually, for any manifold M of nonpositive sectional curvature, the higher homotopy groups $\pi_n(M)$, $n \geq 2$, vanish, and M can be expressed as a quotient space of a Hadamard manifold (the universal covering of M) by a suitable discrete group of isometries of the universal covering isomorphic to $\pi_1(M)$ (see P. Eberlein [34]). The most important case of this being a homogeneous space of a semisimple Lie group of noncompact type.

We have the following basic features in any Hadamard manifold M (see [54]):

1. The exponential map $\exp_p : T_p(M) \to M$ is a diffeomorphism for each $p \in M$.

2. For each pair p, $q \in M$ there exists a unique normal (i.e. unit speed), minimizing geodesic from p to q.

3. For any geodesic triangle in M whose sides are geodesics of length a, b and c, we have the Law of Cosines for manifolds of nonpositive curvature. Namely, $c^2 \geq a^2 + b^2 - 2ab\cos\theta$, where θ is the angle opposite c.

4. The sum of the interior angles of any geodesic triangle is at most π.

5. For any pair of geodesics α, β in M, the function $f(t) = d(\alpha(t), \beta(t))$ is convex.

6. Let C be a compact convex subset of M. Then, for each $p \in M$ there exists a unique point $C(p) \in C$ such that $d(p, C(p)) \leq d(p, q)$ for any $q \in C$. In the Riemannian context, the point $C(p)$ is called the *foot of the perpendicular* from p to C.

Indeed,

Theorem 4.2.6. *Let M be a connected, complete Riemannian manifold of nonpositive sectional curvature. Then, the derivative of the exponential map is length increasing, i.e.*

$$|(\exp_p)_*(v)(w)| \geq |w|$$

for all $p \in M$ and all v, w in $T_p(M)$.

Proof. Let $p \in M$, and let v, and w be two tangent vectors in $T_p(M)$, with $w \neq 0$. Now, consider the geodesic $\gamma(t) = \exp_p(tv)$ and let

$$X(t) := \frac{d}{ds}\Big|_{s=0} \exp_p(t(v + sw))$$

be a vector field along γ. Then $X(0) = 0$, $X(t) = (\exp_p)_*(tv)(tw)$ and

$$\frac{d}{dt}X(0) = \frac{d}{dt}\frac{d}{ds}\exp_p(t(v+sw))\Big|_{s,0} = \frac{d}{ds}\frac{d}{dt}\exp_p(t(v+tw))\Big|_{s,0}$$

$$= \frac{d}{ds}(v+sw) = w \neq 0.$$

Since $X(t)$ is a Jacobi field along γ, by Proposition 4.1.10, it satisfies the equation $X''(t) - R(X, \dot\gamma)\dot\gamma = 0$. We have

$$\frac{d}{dt}|X(t)| = \frac{d}{dt}\left(\langle X(t), X(t)\rangle^{1/2}\right)$$

$$= \frac{1}{2}\langle X(t), X(t)\rangle^{-1/2} 2\langle X(t), X'(t)\rangle = \frac{\langle X(t), X'(t)\rangle}{\langle X(t), X(t)\rangle^{1/2}},$$

and taking the derivative again, we get:

$$\frac{d^2}{dt^2}|X(t)| = \frac{d}{dt}\left(\frac{\langle X(t), X'(t)\rangle}{\langle X(t), X(t)\rangle^{\frac{1}{2}}}\right)$$

$$= \frac{\langle X(t), X''(t)\rangle + \langle X'(t), X'(t)\rangle}{\langle X(t), X(t)\rangle^{\frac{1}{2}}} - \frac{\langle X(t), X'(t)\rangle^2}{\langle X(t), X(t)\rangle^{\frac{3}{2}}}$$

$$= \frac{|X'(t)|^2 + \langle X(t), X''(t)\rangle}{|X(t)|} - \frac{\langle X(t), X'(t)\rangle^2}{|X(t)|^3}$$

$$= \frac{|X(t)|^2 \cdot |X'(t)|^2 - \langle X(t), X'(t)\rangle^2}{|X(t)|^3} - \frac{\langle X(t), X''(t)\rangle}{|X(t)|}.$$

Using the Cauchy-Schwarz inequality,

$$|X(t)|^2 \cdot |X'(t)|^2 - \langle X(t), X'(t)\rangle^2 \geq 0.$$

In addition, since $X(t)$ is a Jacobi field, by Proposition 4.1.10, $X''(t) = -R(X(t), \dot{\gamma})\dot{\gamma}$, and since the curvature is nonpositive, $\langle X(t), X''(t) \rangle \geq 0$. Hence

$$\frac{d^2}{dt^2}\left|X(t)\right| \geq 0.$$

Since the second derivative of the function $|X(t)| - t|w|$ is nonnegative, its first derivative is nondecreasing and since it vanishes at $t = 0$, $|X(t)| - t|w| \geq 0$ for every $t \in (0, +\infty)$. In particular, for $t = 1$,

$$(\exp_p)_*(v)(w) = |X(1)| \geq |w|.$$

$\square$

Theorem 4.2.7. *Let M be a connected complete Riemannian manifold of nonpositive sectional curvature. Then, the exponential map is distance increasing, i.e.*

$$d(\exp_p(\xi_0), \exp_p(\xi_1)) \geq |\xi_0 - \xi_1|$$

for all $p \in M$ and all ξ_0, ξ_1 in $T_p(M)$.

Proof. Let p be any point of M, and ξ_0, ξ_1 be two tangent vectors at p. From Theorem 4.2.6, we know $|(\exp_p)_*(\xi_0)(\xi_1)| \geq |\xi_1|$. Now, consider the two points $\exp_p(\xi_0)$ and $\exp_p(\xi_1)$ of M, and let $\gamma(t)$, $t \in [0, 1]$, be the unique geodesic with endpoints $\gamma(0) = \exp_p(\xi_0)$ and $\gamma(1) = \exp_p(\xi_1)$. Let $\xi(t)$ be the lifting of $\gamma(t)$ in $T_p(M)$ (i.e. $\exp_p(\xi(t)) = \gamma(t)$ for every $t \in [0, 1]$). Then $\xi(0) = \xi_0$ and $\xi(1) = \xi_1$ and

$$L(\gamma) = d(\gamma(0), \gamma(1)) = d(\exp_p(\xi_0), \exp_p(\xi_1)) = \int_0^1 |(\exp_p)_*(\xi(t)(\dot{\xi}(t))dt$$

$$\geq \int_0^1 |\dot{\xi}(t)|dt \geq \left|\int_0^1 \dot{\xi}(t)dt\right| = |\xi_1 - \xi_0|.$$

$\square$

Remark 4.2.8. Actually, more is true. Namely, all these conditions are equivalent: Theorem 4.2.7 implies Theorem 4.2.6, and if the derivative of $\exp_p$ is length increasing, M must have nonpositive sectional curvature everywhere. However, we do not prove this as it is not needed for Cartan's fixed point theorem.

We now turn to the Law of Cosines for any manifold of nonpositive curvature. The following diagram illustrates the situation.

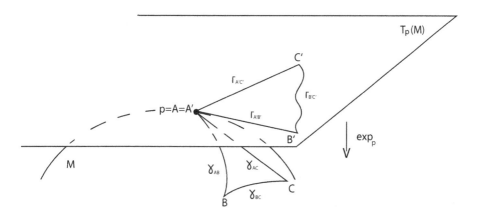

Lemma 4.2.9. *Let M be a Riemannian manifold of nonpositive curvature and U a minimizing convex normal ball in M. Let ABC be a geodesic triangle within U with sides opposite the angles A, B and C respectively, of length a, b and c, (measured through the inner product structure). Then*

1. $b^2 + c^2 - 2bc \cos A \leq a^2$.

2. $A + B + C \leq \pi$.

Proof. Let U be a normal neighborhood of M centered at $p = A$. We know that $\exp_p$ is a diffeomorphism of some neighborhood of $0 \in T_p(M)$ onto U. In addition, by the normal neighborhood theorem, any two points in U are connected by a unique geodesic. Consider the geodesics $\gamma_{AB}, \gamma_{AC}, \gamma_{BC}$ between the respective pairs of points, and let Γ_{AB}, Γ_{BC} be the inverse images of the first and third of these in $T_p(M)$ under $\exp_p$. Note that these are straight lines, but Γ_{BC} may not be a straight line in general.

Let A', B', C' be the points in $T_p(M)$ corresponding to A, B, C respectively. Since geodesics travel at unit speed, $d(A, B) = L(\gamma_{AB}) = L(\Gamma_{AB}) = d(A', B')$ and similarly $d(A, C) = d(A', C')$. Moreover, $d(B, C) = L(\gamma_{BC}) \geq L(\Gamma_{BC}) \leq d(B', C')$, where the first inequality comes from the fact that M has nonpositive curvature and $\exp_p$ increases the lengths of curves (see Theorem 4.2.6). From the Law of Cosines in Euclidean Space we get,

$$d(B', C')^2 = d(A', C')^2 + d(A', B')^2 - 2d(A', B')d(A', C') \cos A.$$

Therefore,

$$d(B', C')^2 \geq d(A', C')^2 + d(A', B')^2 - 2d(A', B')d(A', C') \cos A.$$

Concerning 2), there is in fact an ordinary planar triangle with sides a', b' and c', since these satisfy the appropriate inequalities. The angles A', B', C' of this triangle satisfy $A \leq A', B \leq B', C \leq C'$. Hence, we get $A + B + C \leq \pi$. $\qquad\qquad\qquad\qquad\qquad\qquad\qquad\qquad\qquad\qquad \square$

A final ingredient in the proof of Cartan's fixed point theorem is the following lemma.

Lemma 4.2.10. *Let M be a complete, simply connected Riemannian manifold of nonpositive curvature, $p \in M$ and $\gamma : [0, 1] \to M$ be a geodesic segment not containing p. Then*

$$\frac{d}{dt} d(p, \gamma(t))|_{t=0} = \cos(\alpha),$$

where α denotes the angle formed by γ and the unique geodesic joining p and $\gamma(0)$, at $\gamma(0)$.

Proof. We calculate the derivative of the distance function as the point varies along γ. Assume γ is parameterized by arc length. Let $F_\gamma(t) = d(p, \gamma(t))$. Just as in calculus, we calculate the derivative of the square, $\frac{d}{dt} F_\gamma(t)^2|_{t=0}$. Since $\exp_p : T_p(M) \to M$ is a diffeomorphism. we can replace γ by the lifted curve $r(t)$ in $T_p(M)$ corresponding to γ. By Hopf-Rinow we have $d(p, \gamma(t)) = d(0, r(t)) = |r(t)|$, where the latter distance is in the Euclidean space $T_p(M)$. By the Law of Cosines there,

$$|r(t)|^2 - |r(0)|^2 = |r(t) - r(0)|^2 - 2|r(0)||r(t) - r(0)|\cos(\beta_t),$$

where β_t is the lift of α_t. As $t \to 0$, the first term $|r(t) - r(0)|^2$ is $O(t^2)$ and contributes nothing to the derivative at 0. On the other hand for small t, $|r(t) - r(0)|$ is approximately $t|r'(0)|$. Putting this together we get

$$\frac{d}{dt} d(p, q(t))^2|_{t=0} = -2d(p, q(0))|r'(0)|\cos(\beta_{t=0}).$$

Therefore $F'(0) = -|r'(0)|\cos(\beta)$. Now $|r'(0)|\cos(\beta)$ is the length of the projection of $r'(0)$ onto the line from 0 to $r(0)$. Resolving this vector into components, where the first component is in the radial direction, and the other is orthogonal to this, we see it is $|r'(0)|\cos(\beta)$. In the curve $\gamma(t) = \exp_p(r(t))$, we have $A' = \exp_p(A) \in T_{\gamma(0)}(M)$ is orthogonal to $B' = \exp_p(B)$ and $\exp_p$ in the direction of the geodesic γ. Also, $|A'| = |A|$; so straight lines through the origin are sent to geodesics in M, which travel at constant speed. Note that $|A' + B'| = 1$ by the assumption on γ. In particular, $\cos(\alpha)$ is the length $|A'|$ which we have just seen is $|A| = -|r'(0)|\cos(\beta)$. $\square$

We are now in a position to prove Cartan's Theorem 4.0.1.

Proof. Let $d\mu$ be normalized Haar measure on C and define $J(q) = \int_C d^2(q, cp)d\mu(c)$. This is a continuous function $M \to \mathbb{R}$ which has a minimum, because $J(q) > J(p)$ for q outside some compact set containing p. Suppose the minimum occurs at q_0. We show the minimum is unique, which will imply that it is a fixed point for C.

J is continuously differentiable. Indeed, let q_t be a curve. Then $d^2(q_t, kp)$ can be computed as above, when $kp \neq q_t$. When they are equal, it is still differentiable with zero derivative because of the d^2. So, now take q_t to be a geodesic joining the minimal point q_0 to some other point q_1. Then

$$\frac{d}{dt}J(q_t)|_{t=0} = \int_K \frac{d}{dt}d^2(q_t, kp)|_{t=0}\, d\mu(k) = 0.$$

Hence,

$$\int_K d(q_0, kp)\cos \alpha\, d\mu(k) = 0,$$

where α is an appropriate angle. When $q_0 = kp$, this α is not well-defined, but $d(q_0, kp) = 0$, so we are still ok here. Now,

$$\int_K d^2(q_1, k.p)d\mu \geq \int_K \left(d^2(q_0, kp) + d^2(q_0, q_1) - 2d(q_0, kp)\cos \alpha\right) d\mu(k).$$

This is because of the cosine inequality. But the cosine part vanishes, so this is strictly greater than $J(q_0)$. In particular, since q_1 was arbitrary, q_0 was a global minimum for J and is thus a fixed point. $\square$

The conjugacy of maximal compact subgroups of semisimple Lie group of noncompact type is due to E. Cartan and follows from his fixed point theorem. Actually, this result holds for an arbitrary connected Lie group and is due to K. Iwasawa, and the case of a finite number of components to G. D. Mostow. In this more general context see [57]. The proof in [57] is similar to the one given here, but rather than involve the differential geometry itself, uses a convexity argument and a function which mimics the metric. Of course, similarly to applying Borel's fixed point theorem to prove Lie's theorem (see Chapter 6), our approach requires knowing $X = G/K$(where G is a connected semisimple Lie group of noncompact type and K is a maximal compact subgroup)is a Hadamard manifold (see [1]).

Corollary 4.2.11. *The maximal compact subgroups in a connected semisimple Lie group G of noncompact type are all conjugate. Any compact subgroup of G is contained in a maximal one.*

Proof. Let C be a compact subgroup of G. By the fixed point theorem 4.0.1, there is a C-fixed point $x_0 \in X = G/K$, where K is a maximal compact subgroup of G. Thus $C \subseteq \mathrm{Stab}_G(x_0)$. But this action is transitive so $\mathrm{Stab}_G(x_0) = gKg^{-1}$ for some $g \in G$. Since K is a maximal compact evidently so is the conjugate gKg^{-1}. This proves the second statement. If C were itself maximal, then $C = gKg^{-1}$. □

4.3 Fixed Point Theorems for Compact Manifolds

We first mention the following interesting result of Preissmann concerning *compact* manifolds of negative curvature. These usually arise in the form $\Gamma\backslash G/K$, where G and K are as above and Γ is a cocompact lattice in G.

Theorem 4.3.1. (Preissmann Theorem, 1943). *If (M, g) is a compact manifold of negative curvature, then any Abelian subgroup of the fundamental group is cyclic. In particular, no compact product manifold $M \times N$ admits a metric of negative curvature.*

The reader can find a proof of this in [85], p. 167. This last statement generalizes and explains the very intuitive fact that $S^1 \times S^1$ has points of negative, zero and positive curvature.

We now give some applications of the same general techniques as above, but this time applied to the case of compact Riemannian manifolds of positive curvature.

Proposition 4.3.2. *Let $c : [a, b] \to M$ be a smooth curve in M and $E(c)$ denote the energy of c. Then*

$$E'(0) = \langle c'(b, 0), \dot{c}(b, 0) \rangle - \langle c'(a, 0), \dot{c}(a, 0) \rangle - \int_a^b \Big\langle \frac{\partial c}{\partial s}, \nabla_{\frac{\partial}{\partial t}} \frac{\partial c}{\partial t}(t, s) \Big\rangle dt.$$

Proof. We know

$$E(s) = \frac{1}{2} \int_a^b \Big\langle \frac{\partial c}{\partial t}(t, s), \frac{\partial c}{\partial t}(t, s) \Big\rangle dt.$$

Therefore,

$$\frac{d}{ds}E(s) = \frac{1}{2}\int_a^b \frac{\partial}{\partial s}\Big\langle \frac{\partial c}{\partial t}(t,s), \frac{\partial c}{\partial t}(t,s)\Big\rangle dt$$

$$= \int_a^b \Big\langle \nabla_{\frac{\partial}{\partial t}} \frac{\partial c}{\partial t}(t,s), \frac{\partial c}{\partial t}(t,s)\Big\rangle dt,$$

and, since ∇ preserves the metric, this is equal to

$$= \int_a^b \Big\langle \nabla_{\frac{\partial}{\partial s}} \frac{\partial c}{\partial t}(t,s), \frac{\partial c}{\partial t}(t,s)\Big\rangle dt.$$

Taking into consideration that ∇ is torsion free, we get

$$= \int_a^b \Big\langle \nabla_{\frac{\partial}{\partial t}} \frac{\partial c}{\partial s}(t,s), \frac{\partial c}{\partial t}(t,s)\Big\rangle dt$$

$$= \int_a^b \Big(\frac{\partial}{\partial t}\Big\langle \frac{\partial c}{\partial s}(t,s), \frac{\partial c}{\partial t}(t,s)\Big\rangle - \Big\langle \frac{\partial c}{\partial s}, \nabla_{\frac{\partial}{\partial t}} \frac{\partial c}{\partial t}(t,s)\Big\rangle \Big) dt$$

$$= \Big\langle \frac{\partial c}{\partial s}, \frac{\partial c}{\partial t}\Big\rangle\Big|_{t=a}^{t=b} - \int_a^b \Big\langle \frac{\partial c}{\partial s}, \nabla_{\frac{\partial}{\partial t}} \frac{\partial c}{\partial t}(t,s)\Big\rangle dt.$$

$\square$

We now compute the *second variation* of the energy function. To do this, consider the geodesic $\gamma : [0,1] \to M$ joining points p and q of M, (i.e. a critical point of the energy function $E : \Omega(M,p,q) \to \mathbb{R}$) and two vector fields W_1, W_2 along $\gamma(t)$. Let F be a 2-parameter differentiable variation of γ. That is, a smooth function $F : U_{(0,0)} \times [0,1] \to M$ such that $(s_1, s_2, t) \mapsto F(s_1, s_2, t)$, where $U_{(0,0)}$ is a neighborhood of $(0,0) \in \mathbb{R}^2$, and satisfies:

$$F(0,0,t) = \gamma(t), \quad \frac{\partial}{\partial s_1}F(0,0,t) = W_1(t), \quad \frac{\partial}{\partial s_2}F(0,0,t) = W_2(t).$$

Then, we define (see [70]) the *Hessian* $E_{\star\star}(W_1, W_2)$ by setting

$$E_{\star\star}(W_1, W_2) := \frac{\partial^2}{\partial s_1 \partial s_2}E\Big(F_t(s_1,s_2)\Big)\Big|_{(0,0)}.$$

Proposition 4.3.3. (Second Variation of the Energy). *Let M be a Riemannian manifold, $\gamma : [0,1] \to M$ be a geodesic that joins the points p and q of M and let F be a 2-parameter variation of γ as above. Then, $E_{\star\star}(W_1, W_2)$ is equal to*

$$-\sum_t \left\langle W_2(t), \Delta_t \frac{DW_1}{dt} \right\rangle - \int_0^1 \left\langle W_2, \frac{D^2 W_1}{dt^2} + R(V, W_1)V \right\rangle dt,$$

where $V = \frac{d\gamma}{dt}$ and

$$\Delta_t \frac{DW_1}{dt} = \frac{DW_1}{dt}(t^+) - \frac{DW_1}{dt}(t^-)$$

denotes the jump of $\frac{DW_1}{dt}$ at one of its finitely many points of discontinuity in $(0,1)$.

Proof. Applying the first variation formula to the geodesic γ we get

$$\frac{1}{2} \frac{\partial E}{\partial s_1} = -\sum_t \left\langle \frac{\partial F}{\partial s_2}, \Delta_t \frac{\partial F}{\partial t} \right\rangle - \int_0^1 \left\langle \frac{\partial F}{\partial s_2}, \frac{D}{dt} \frac{\partial F}{\partial t} \right\rangle dt.$$

Hence, we get

$$\begin{aligned}
\frac{1}{2} \frac{\partial^2}{\partial s_1 \partial s_2} = &-\sum_t \left\langle \frac{D}{ds_1} \frac{\partial F}{\partial s \partial_2}, \Delta_t \frac{\partial F}{\partial t} \right\rangle - \sum_t \left\langle \frac{\partial F}{\partial s_2}, \frac{D}{\partial s_1} \Delta_t \frac{\partial F}{\partial t} \right\rangle \\
&- \int_0^1 \left\langle \frac{D}{ds_1} \frac{\partial F}{\partial s_2}, \frac{D}{dt} \frac{\partial F}{\partial t} \right\rangle dt - \int_0^1 \left\langle \frac{\partial F}{\partial s_2}, \frac{D}{ds_1} \frac{D}{dt} \frac{\partial F}{\partial t} \right\rangle dt.
\end{aligned}$$

Since $\gamma(t) = F(0,0,t)$ is an unbroken geodesic, evaluating the above expression at $(0,0)$, we get

$$\Delta_t \frac{\partial F}{\partial t} = 0, \quad \frac{D}{dt} \frac{\partial F}{\partial t} = 0,$$

thus the first and third terms vanish, and by rearranging the second one, we get

$$\frac{1}{2} \frac{\partial^2 E}{\partial s_1 \partial s_2} \Big|_{(0,0)} = -\sum_t \left\langle W_2, \Delta_t \frac{D}{dt} W_1 \right\rangle - \int_0^1 \left\langle W_2, \frac{D}{ds_1} \frac{D}{dt} V \right\rangle dt.$$

Now we have,

$$\frac{D}{ds_1}\frac{D}{dt}V - \frac{D}{dt}\frac{D}{ds_1}V = R\left(\frac{\partial F}{\partial t}, \frac{\partial F}{\partial s_1}\right)V = R(V, W_1)V,$$

which when combined with the identity

$$\frac{D}{ds_1}V = \frac{D}{dt}\frac{\partial F}{\partial s_1} = \frac{D}{dt}W_1$$

gives us

$$\frac{D}{ds_1}\frac{D}{dt}V = \frac{D^2 W_1}{dt^2} + R(V, W_1)V.$$

By replacing this in the expression we get the conclusion.   □

We now prove Weinstein's fixed point theorem concerning compact manifolds of positive sectional curvature.

Theorem 4.3.4. *Let M be a compact oriented n-dimensional Riemannian manifold of positive sectional curvature and $f : M \to M$ be an isometry. In addition, suppose that f preserves orientation when n is even, and reverses it when n is odd. Then f has a fixed point[7].*

To prove this we will need the following fact of Linear Algebra: The case $n = 3$ was familiar to Euler.

Lemma 4.3.5. *Let $T : V \to V$ be an orthogonal linear transformation of a real vector space V of dimension n. If $\det(T) = (-1)^{n+1}$, then T has a nontrivial fixed point.*

Proof. First, we note that since T is orthogonal all its *real* eigenvalues are ± 1. Now, if n is even, $\det(T) = -1$ and since the product of the complex eigenvalues is never negative (because they are complex conjugates), the characteristic polynomial $\det(T - \lambda I)$ must have as zeros at least a pair of real numbers with product -1; hence one of them is 1.

---

[7]Actually, in [101] Weinstein proved this theorem for conformal maps and not only isometries. Indeed, there is a conjecture that this may be true for arbitrary diffeomorphisms.

When n is odd the characteristic polynomial must have an odd number of real roots. Again, since the product of complex eigenvalues is
nonnegative, and by assumption $\det(T) = 1$, T must have at least one
eigenvalue equal to 1.   □

Proof of Theorem 4.3.4.

Proof. Suppose that the isometry f has no fixed point, i.e. $f(p) \neq p$ for
every $p \in M$. Since M is a compact manifold, the function $g : M \to \mathbb{R}$
given by $g(p) = d(p, f(p))$ achieves its minimum at some point $p_0 \in M$
and $g(p) > 0$ everywhere on M. Since M is complete, the Hopf-Rinow
theorem tells us there is a minimizing geodesic $\gamma : [0.l] \to M$ such that
$\gamma(0) = p$ and $\gamma(l) = f(p)$. We want to show the curve $\gamma + f \circ \gamma$ forms
a forward geodesic. For this, we consider the forward geodesic $f \circ \gamma$
which connects $f(p)$ with $f^2(p)$. Take a point $q = \gamma(t)$, $0 < t < l$,
on the geodesic γ lying between p and $f(p)$. Since f is an isometry
$d(p, q) = d(f(p), f(q))$, for all p and q. By the triangle inequality

$$d(q, f(q)) \leq d(q, f(p)) + d(f(p), f(q)) = d(q, f(p)) + d(p, q) = d(p, f(p)).$$

Since $d(p, f(p))$ is a minimum, $d(q, f(q)) = d(q, f(p)) + d(f(p), f(q))$,
which means that the curve formed by γ and $f \circ \gamma$ (i.e. the curve
$\gamma + f \circ \gamma$) is a geodesic. This implies that

$$(f \circ \gamma)'(0) = \gamma'(l).$$

Now, consider the map $\widetilde{T} := P \circ df_p : T_p(M) \to T_p(M)$ where P is the
parallel transport along γ from $f(p) = \gamma(1)$ to p. $\widetilde{T}$ leaves $\gamma'(0)$ fixed
since

$$\widetilde{T}(\gamma'(0)) = P \circ df_p(\gamma'(0)) = P((f \circ \gamma)'(0)) = P(\gamma(1)) = \gamma'(0).$$

Consider the $n - 1$ dimensional subspace W of $T_p(M)$ defined by

$$W = \{\gamma'(0)\}^{\perp}$$

and the restriction of T to W. Since P is an isometry preserving orientation,

$$\det(T) = \det(\widetilde{T}) = \det(P \circ df_p) = (-1)^n,$$

and by Lemma 4.3.5 T leaves a vector invariant. Let $V(t)$ be a unit parallel vector field along γ such that for each t, $V(t) \in \{\gamma'(t)\}^\perp$ and $V(0)$ is invariant by T. Consider the variation

$$\gamma_s(t) = \exp_{\gamma(t)}(sV(t)).$$

Then γ_s connects $\gamma_s(0) = \exp_p(sV(0))$ to $f(\gamma_s(0))$. This implies there is a curve c in the variation whose length is smaller than the length of γ, contradicting the fact that $d(p, f(p))$ is a minimum. □

We note that the assumption in Weinstein's theorem concerning the preservation of orientation is crucial. For let $f : S^2 \to S^2$ be the antipodal map of the 2-sphere, S^2. Then $f = -Id$. Obviously, f reverses orientation and evidently has no fixed point.

As corollary we get Synge's theorem.

Corollary 4.3.6. *Let M be a compact n-dimensional Riemannian manifold of positive curvature. If M is oriented and of even dimension, then it is simply connected. If M is of odd dimension it is orientable.*

Of course, the 1-dimensional torus shows the hypothesis of even dimensionality here is essential in the first statement. It also tells us, as we saw earlier, an even dimensional torus cannot support any Riemannian metric of positive curvature.

Proof. Suppose not. Then there exists a nontrivial homotopy class of loops, say $[d]$. Since M is compact we claim in $[d]$ there is a smooth closed geodesic γ (where $\gamma \sim d$) of minimal length among all such loops in $[d]$. To see this let l be the infimum of the lengths of all curves in $[d]$. Consider a subdivision of $[0,1]$, as $0 \le t_0 < \cdots < t_n = 1$, such that $t_{i+1} - t_i < \frac{i(M)}{l}$ for all i, where $i(M)$ is the injectivity radius of M. Fix a point $p \in \gamma$ and consider parallel translation around γ to obtain a linear isometry of the even-dimensional vector space $\iota : T_p(M) \to T_p(M)$. Because γ is a geodesic, $i(\gamma') = \gamma'$. Since M is orientable, $i : (\gamma')^\perp \to (\gamma')^\perp$ is an orthogonal transformation with determinant 1 of an odd-dimensional real inner product space. As above 1 must be an eigenvalue. Hence there exists a smooth parallel unit vector field W along γ normal to γ'. The second variation formula of Theorem 4.3.3 implies

$$\frac{d^2}{ds^2}|_{s=0}E(\gamma_s) = -\int_a^b \langle R(W,T)T, W \rangle \, dt < 0,$$

because the sectional curvatures are positive. This contradicts the fact that γ has minimal length in its homotopy class. Hence M is simply connected.

For the second statement, suppose M is not orientable. Then, it must have an orientable double cover, $\widetilde{M}$ which is also compact and has an induced Riemannian metric from M, also of positive curvature. Now, let $\varphi : \widetilde{M} \to \widetilde{M}$ be a covering transformation with $\varphi \neq I$. Since M is not orientable φ is an isometry which reverses the orientation of $\widetilde{M}$, and since M is of odd dimension, by Theorem 4.3.4 it must have a fixed point. But, by Corollary 6.2 of Massey [68], the group of its covering transformations operates on M without fixed points, i.e. if $\varphi \neq I$, then φ has no fixed point. Thus $\varphi = I_M$, a contradiction. $\square$

Chapter 5

Fixed Points of Volume Preserving Maps

It should be remarked that here we are in a somewhat tricky environment. For a smooth manifold M, the Myers-Steenrod theorem [26] tells us that the subgroup of isometries within $\operatorname{Diff}(M)$ is always a Lie group. However, even when M is compact and connected, the subgroup of all volume preserving diffeomorphisms is too big to be a Lie group.

5.1 The Poincaré Recurrence Theorem

This result, while not a fixed point theorem, is an approximate fixed point theorem. Both its statement and method is of great importance in dynamical systems and elsewhere.

Theorem 5.1.1. *Let (X, μ) be a finite measure space, T a measure preserving map of X and U be a subset of X of positive measure. Then for almost all $u \in U$, $T^n(u) \in U$ for infinitely many positive integers n.*

Of particular importance is the case when X is a locally compact space, μ is a regular measure and U is a nonempty neighborhood in X. Under these circumstances Theorem 5.1.1 is a kind of a fixed point theorem. It says that for *almost all $u \in U$*, many powers $T^n(u)$, if not equal to u, are at least near u. That is when the total measure is finite, under iteration most u's eventually return arbitrarily near their initial position. In particular, for a dense set D in U, $U \subseteq \overline{\{T^n(D) : n \geq 1\}}$.

We now make the following definition.

Definition 5.1.2. Let $f : X \to X$ be a continuous map of a topological space. A point $x \in X$ is said to be *nonwandering* if for any neighborhood U of x there is a positive integer k so that $f^k(U) \cap U$ is nonempty, otherwise we say x is *wandering*. We denote by $\Omega(f)$ the nonwandering points of f.

A direct consequence of Theorem 5.1.1 is:

Corollary 5.1.3. *If $T : X \to X$ is a volume preserving homeomorphism, then X has no wandering points.*

We note that the assumption that the measure is finite in the recurrence theorem is crucial. For taking $X = \mathbb{R}$ with Lebesgue measure and for T the shift operator $x \mapsto x + 1$, T preserves measure, but no point $x \in \mathbb{R}$ is recurrent since all points go to infinity. We now prove Theorem 5.1.1.

Proof. Let N be a fixed positive integer and

$$U_N = \{u \in U : T^n(u) \in X \setminus U \text{ for all } n \geq N\}.$$

We will show $\mu(U_N) = 0$. Suppose $\mu(U_N) > 0$. For positive integers r, consider $T^{rN}(U_N)$. These sets all have measure $\mu(U_N) > 0$. Therefore they cannot be disjoint. For if they were, by countable additivity we would conclude $\mu(X) \geq \mu(\cup_{r=1}^{\infty} T^{rN}(U_N)) = \infty$, a contradiction. Hence there must be $r_1 < r_2$ so that $T^{r_1 N}(U_N) \cap T^{r_2 N}(U_N)$ is nonempty. That is there is u_1 and $u_2 \in U_N$ for which $T^{r_1 N}(u_1) = T^{r_2 N}(u_2)$. That is, $T^{(r_2-r_1)N}(u_2) = u_1$. Since u_1 and u_2 are both in $U_N \subseteq U$ this contradicts the definition of U_N and proves $\mu(U_N)$ must be 0. This means that for fixed N, almost all $u \in U$ satisfy $T^n(u) \in U$ for some $n \geq N$. Since N is arbitrary and there are countably many of them the conclusion follows. $\square$

Some remarks:

The Poincaré Recurrence theorem gives rise to the following paradox: In classical physics the state of a system of N particles is described
completely by a point in the *phase space* X which evolves according to
Hamilton's equations. For example, let X include all possible states
(the finite) set of molecules in a box divided in two by a wall. Here the
σ-algebra, $\mathcal{A}$, is the collection of all observable states of the system and
$\mu(A)$ is the probability of observing the state A. Let T denote the time
evolution of the system. It is reasonable to expect that if the system
is in equilibrium, T preserves μ. That is, the probability of observing
a certain state is independent of time. Thus, we seem to be in a situation where the Poincaré Recurrence theorem should apply. Suppose in
the initial state all the particles in the box are in the left half. Letting
U denote the neighborhood which is the left side, Poincaré Recurrence
tells us at some future time *almost all* the molecules will return to that
half of the box. This contradicts the 2nd law of Thermodynamics.

A possible explanation of this paradox, is that the theorem says
that this will occur, but of course, it does not say anything about when.
Suppose $T^{k_i} \in A$ for $0 = k_0 < k_1 \ldots k_i < \ldots$, and only for those
values. Then k_1 is called the *recurrence time*. The differences between
consecutive recurrent times $k_i - k_{i-1}$ is denoted by R_i. A consequence
of the ergodic theorem is that the average of these R_i's is inversely
proportional to the measure of A, That is,

$$\lim_{n \to \infty} \frac{R_1 + \cdots + R_n}{n} = \frac{\mu(X)}{\mu(A)}.$$

Thus, as is intuitively obvious, the smaller A is, the longer it takes
(on average) to return to it. Since in our situation the phase space is
huge (indeed under normal conditions, the number of gas molecules in
a volume of $1 \ cm^3$ is 10^{23}). Since the observables corresponding to all
molecules in the left half of the box has extremely small measure in
this large space, the time it will take for this configuration to recur is
huge, probably longer than the age of the universe (which is believed to
be less than 10^{18} seconds old) and so it is not possible to observe this.

However, another possibility is that the global solution to Hamilton's equations is just not a measurable function!

5.2   Symplectic Geometry and its Fixed Point Theorems

5.2.1   Introduction to Symplectic Geometry

We begin with Symplectic[1] vector spaces. Let V be a finite-dimensional, real vector space with dual, $V^\star$. The space $\bigwedge^2 V^\star$ can be identified with the space of all skew-symmetric $(\omega(v,w) = -\omega(w,v))$ bilinear forms $\omega : V \times V \to \mathbb{R}$.

Definition 5.2.1. (V, ω) is called a *symplectic vector space* if ω is non-degenerate,

$$\{v \in V \ : \ \omega(v, w) = 0 \ \text{for all} \ w \in V\} = (0).$$

We say that the two symplectic vector spaces (V_1, ω_1) and (V_2, ω_2) are *symplectomorphic* if there is an isomorphism $A : V_1 \to V_2$ such that $A^\star \omega_2 = \omega_1$. The set of all symplectomorphisms is a Lie group (as a closed subgroup of $\mathrm{GL}(V)$), and it is denoted by $\mathrm{Sp}(V)$ (sometimes written $\mathrm{Sp}(n, \mathbb{R})$).

Example 5.2.2. Let V be a vector space, and $V^\star$ its dual. Then $V \oplus V^\star$ becomes a symplectic space with ω given by

$$\omega\Big((v_1, v_1^\star), (v_2, v_2^\star)\Big) := v_1^\star(v_2) - v_2^\star(v_1).$$

Definition 5.2.3. A *complex structure* on a real vector space V is an automorphism $J : V \to V$ satisfying $J^2 = -\mathrm{I}_V$. In addition, a complex structure J on a symplectic vector space (V, ω) is called ω-*compatible* if $g(v, w) = \omega(v, Jw)$ is a positive definite inner product.

---

[1]The terminology *symplectic* was first introduced by Hermann Weyl in [102] in 1939, p. 165. Its meaning in Greek is interlacing.

In this case J is a symplectomorphism. Indeed,

$$(J^\star\omega)(v,w) = \omega(Jv, Jw) = g(Jv, w) = g(w, Jv)$$
$$= \omega(w, J^2v) = -\omega(w, v) = \omega(v, w).$$

Example 5.2.4. Take $V = \mathbb{R}^{2n}$ with basis $\{e_1, \ldots, e_n, f_1, \ldots, f_n\}$. Then we define the standard symplectic structure on $\mathbb{R}^{2n}$ by setting:

$$\omega(e_i, e_j) = 0, \quad \omega(f_i, f_j) = 0, \quad \omega(e_i, f_j) = \delta_{ij} = -\omega(f_j, e_i).$$

Here we get symplectomorphisms as follows: $A(e_j) = f_j$, $A(f_j) = -e_j$, or $A(e_j) = e_j + f_j$, $A(f_j) = f_j$. Now, a compatible complex structure J is given by $Je_i = f_i$, $Jf_i = -e_i$. With this we identify $(\mathbb{R}^{2n}, \omega, J)$ with $\mathbb{C}^n$.

Let A^t denote the transpose of an endomorphism A with respect to g as above. Then

Lemma 5.2.5. *An automorphism $A \in \mathrm{GL}(V)$ is in $\mathrm{Sp}(V)$ if and only if*

$$A^t = JA^{-1}J^{-1}.$$

Proof. We have $A \in \mathrm{Sp}(V)$ if and only if for all v, $w \in V$, $\omega(Av, Aw) = \omega(v, w)$, or equivalently $g(JAv, Aw) = g(Jv, w)$, i.e. $A^t JA = J$. $\square$

For the space $\mathbb{R}^{2n}$ with the standard symplectic structure, J is given by a block-matrix

$$J = \begin{pmatrix} 0 & I \\ -I & 0 \end{pmatrix}.$$

In particular, for $n = 1$, $\mathrm{Sp}(\mathbb{R}^2) = \mathrm{SL}(2, \mathbb{R})$.

We now turn to Symplectic Manifolds.

Let M be a smooth real manifold. We call a *symplectic form* on M a nondegenerate, closed 2-form ω, where here closed means $d\omega = 0$ and nondegenerate means that for each $p \in M$ and $X \in T_p(M)$,

$$\omega(X, Y)(p) = 0, \quad \text{implies } Y = 0.$$

This means for any $p \in M$, the map

$$T_p(M) \to T_p^\star(M),$$

defined by

$$X_p \mapsto (\iota_X \omega)(p) := \omega(X, \cdot)(p),$$

is an isomorphism. In other words, there is a $1 : 1$ correspondence between the 1-forms $\iota_{(}X)\omega \in T^\star(M)$ and the vector fields $X \in T(M)$. The pair (M, ω) is called a *symplectic manifold*. Symplectic manifolds are even dimensional and if $\dim(M) = 2n$, $\omega^n \neq 0$ is a volume form. Hence M is oriented. If M is a symplectic manifold $H^2(M, \mathbb{R}) \neq 0$. In fact more generally we have,

Proposition 5.2.6. *Let M be a compact symplectic manifold of real dimension $2n$. Then $H^{2k}(M, R) \neq (0)$ for all k, $1 \leq k \leq n$.*

Proof. If (M, ω) is a compact symplectic manifold with $\dim M = 2n$, the de Rham cohomology class $[\omega^n] \in H^{2n}(M, \mathbb{R})$ must be nonzero by Stokes' theorem. Therefore, the class $[\omega]$ must be nonzero, as well as its powers $[\omega]^k = [\omega^k] \neq 0$. (Exact symplectic forms can only exist on noncompact manifolds.) $\square$

Since the sphere S^2 is an orientable surface it must have a symplectic structure (given by the standard area form). However, for $n > 1$, the compact manifold S^{2n} has $H^2(S^{2n}, \mathbb{R}) = (0)$ and so it cannot be symplectic whereas as we shall see the torus T^{2n} always has a symplectic structure induced from the standard symplectic structure of $\mathbb{R}^{2n}$ (which is preserved by translations).

Example 5.2.7. The Euclidean space $\mathbb{R}^{2n}$ carries the standard symplectic structure, given by the 2-form $\omega = \sum_i dx_i \wedge dy_i$. This form is exact, since $\omega = d\theta$, where

$$\theta = \sum_{i=0}^{n} y_i dx_i.$$

In addition, if v, $w \in \mathbb{R}^{2n}$, we have

$$
\begin{aligned}
\omega(v,w) &= \sum_{i=1}^{n} \Big(dx_i \otimes dy_i(v,w) - dy_i \otimes dx_i(v,w) \Big) \\
&= \sum_{i=1}^{n} (v_i w_{n+i} - v_{n+i} w_i) \\
&= v^t \cdot J \cdot w = \langle J \cdot v, w \rangle,
\end{aligned}
$$

where $\langle \cdot, \cdot \rangle$ is the standard inner product on $\mathbb{R}^{2n}$. Similarly for the space $\mathbb{C}^n$, but now taking $\omega = \sum_i dz_i \wedge d\bar{z}_i$. Also, every orientable 2-dimensional surface is symplectic. For this, take ω to be any nonvanishing volume form.

We note that for any smooth manifold X, the cotangent bundle $M \equiv T^\star(X)$, is always a symplectic manifold. Indeed, given a chart $(U, x_1, ..., x_n)$ of X, a basis of $T_p^\star(X)$ is given by $dx_1, ..., dx_n$ and every $\xi \in T^\star(X)$ can be written as

$$
\xi = \sum_i \xi_i dx_i.
$$

This gives us a map,

$$
T^\star(X)|_U \to R^{2n}, \quad (x, \xi) \mapsto (x_1, \ldots, x_n, \xi_1, \ldots, \xi_n).
$$

Let λ be the *Liouville form* defined by $\sum_i \xi_i dx_i$ on each chart. It is well defined as a 1-form on M, and

$$
\omega = d\lambda = \sum_i d\xi_i dx_i,
$$

is the desired symplectic form.

Definition 5.2.8. A submanifold L of the symplectic manifold (M, ω) is called *Lagrangian* if $\omega|_L = 0$ and $\dim L = \frac{1}{2} \dim M$.

We remark that for a manifold X, the 0-section $X \to T^\star(X) \equiv M$ is a Lagrangian submanifold. Furthermore, sections of $T^\star(X)$ are the graphs $X_\alpha = \{(x, \alpha(x)) \mid x \in X\} \subset T^\star(X)$ of 1-forms $\alpha \in \Omega^1(X, \mathbb{R})$. Such a graph is Lagrangian if and only if $d\alpha = 0$, since setting $\iota_\alpha(x) = (x, \alpha(x))$,

$$\iota_\alpha \lambda = \alpha, \; \Rightarrow \; \iota_\alpha^\star(\omega) = \iota_\alpha^\star(d\lambda) = d\iota_\alpha^\star \lambda = d\alpha.$$

Definition 5.2.9. Given two symplectic manifolds (M, ω_1), and (N, ω_2), a *symplectomorphism* is a diffeomorphism $f : M \to N$,

$$f^\star \circ \omega_2 = \omega_1.$$

We denote the group of symplectomorphisms of M by $\mathrm{Symp}(M, \omega)$. *As the reader will see below, this is not a Lie group.*

For the 2-sphere, S^2, $\mathrm{Symp}(S^2)$ is the group of area and orientation preserving diffeomorphisms. This is much larger than the group of isometries (which is a Lie group).

Now let $\phi : (M_1, \omega_1) \to (M_2, \omega_2)$ be a diffeomorphism. We want to know whether ϕ is a symplectomorphism as well, i.e. whether $\phi^\star \omega_2 = \omega_1$. For this, consider the graph $\Gamma_\phi \subset M \equiv M_1 \times M_2$ and let π_i be the projection onto M_i. The space M has a symplectic structure via

$$\omega = \pi_1^\star \omega_1 - \pi_2^\star \omega_2.$$

Then we have,

Proposition 5.2.10. ϕ *is a symplectomorphism if and only if Γ_ϕ is Lagrangian.*

Proof. Γ_ϕ is the image of the embedding $\gamma : M_1 \to M_1 \times M_2$ so that $p \mapsto (p, \phi(p))$ and

$$\gamma^\star \omega = \gamma^\star \pi_1^\star \omega_1 - \gamma^\star \pi_2^\star \omega_2 = \omega_1 - \phi^\star \omega_2$$

is equal to 0 if and only if Γ_ϕ is Lagrangian. □

Definition 5.2.11. Let $H : M \to \mathbb{R}$ be a smooth function. This gives rise to a unique vector field X_H satisfying,

$$i_{X_H}\omega = \omega(X_H, \cdot) = dH,$$

under the identification $T(M) \equiv T^\star(M)$ defined by the symplectic form ω. Thus, for each vector field Y on M,

$$dH(Y) = \omega(X_H, Y).$$

We call X_H the *Hamiltonian vector field* and H its *Hamiltonian function.*

Here is an example of a Hamiltonian vector field and flow:

Let $(\mathbb{R}^2, \omega = dx \wedge dy)$ be the symplectic manifold and consider the Hamiltonian function,

$$H : \mathbb{R}^2 \to \mathbb{R}, \quad H(x,y) = \frac{1}{2}(x^2 + y^2).$$

Then, $dH = xdx + ydy$. So

$$\iota\left(x\frac{\partial}{\partial y} - y\frac{\partial}{\partial x}\right)\omega = -xdx - ydy = -dH.$$

Therefore the Hamiltonian vector field is

$$X_H = x\frac{\partial}{\partial y} - y\frac{\partial}{\partial x}.$$

Now, let $\phi_t : M \to M$ be the family of diffeomorphisms generated by the flow of X_H, i.e., $\phi_0 = I$ and $\frac{d}{dt}\phi_t(x) = X_H(\phi_t(x))$. Then each ϕ_t is a symplectomorphism, i.e. $\phi_t^\star\omega = \omega$. Indeed, in addition,

$$\frac{d}{dt}\phi_t^\star\omega = \phi_t^\star(di_{X_H}\omega + i_{X_H}d\omega) = \phi_t^\star(d(dH) + 0) = 0.$$

This shows that the group of symplectomorphisms $\text{Symp}(M, \omega)$ is infinite dimensional and its Lie algebra contains all Hamiltonian vector fields. As noted earlier this is in contrast with the case of Riemannian metrics, where isometry groups are *always Lie groups.*

We mention two important results, Darboux's theorem and Moser's stability theorem. The first shows all symplectic forms are locally equivalent, in sharp contrast to the case of a Riemannian metric, where curvature provides a local invariant, while the second shows that exact deformations of a symplectic structures are trivial.

Definition 5.2.12. A vector field X is said to be *symplectic* if its flow $\phi_X(t)$ consists of symplectomorphisms: that is,

$$\phi_X^\star(t)\omega = \omega, \quad \text{for all } t.$$

Let $\mathcal{L}_X$ denote the Lie derivative. Since

$$\frac{d}{dt}\phi_X^\star(t)\omega = \phi_X^\star(t)(\mathcal{L}_X\omega),$$

X is symplectic if and only if $\mathcal{L}_X\omega = 0$.
When is this zero?

$$\mathcal{L}_X\omega = i_X d\omega + d(i_X\omega) = d(i_X\omega),$$

shows that X is symplectic exactly when the corresponding 1-form $\alpha = i_X\omega$ is closed. Since every manifold supports many closed 1-forms, the group $\text{Symp}(M,\omega)$ of all symplectomorphisms is infinite dimensional.

Therefore, the vector field X is called symplectic when $i_X\omega$ is closed, and Hamiltonian when $i_X\omega$ is exact. Now, the group $\text{Symp}(M,\omega)$ has a normal subgroup $\text{Ham}(M,\omega)$ that corresponds to the exact 1-forms $\alpha = dH$. By definition, $\phi \in \text{Ham}(M,\omega)$ if it is the endpoint of a path $\phi(t)$, $t \in [0,1]$, starting at the identity $\phi_0 = \text{I}$ that is tangent to a family of vector fields X_t for which $i_{X_t}\omega$ is exact for all such t. In this case there is a time independent function $H_t : M \to \mathbb{R}$ (called the *generating Hamiltonian*) for which $i_{X_t}\omega = dH_t$ for all t.

When the first Betti number $b_1 = \dim H^1(M,\mathbb{R})$ of M vanishes, $\text{Ham}(M,\omega)$ is simply the identity component $\text{Symp}_0(M,\omega)$ of the symplectomorphism group. In general, there is a short exact sequence

$$(0) \to \text{Ham}(M,\omega) \to \text{Symp}_0(M,\omega) \to H^1(M,\mathbb{R})/\Gamma_\omega \to (0),$$

where the flux group Γ_ω is a subgroup of $H^1(M, \mathbb{R})$. (In the case of the torus T^2 with a symplectic form $dx \wedge dy$ of total area 1, the group Γ_ω is actually $H^1(M, \mathbb{Z})$.) The family of rotations $R_t : (x, y) \mapsto (x+t, y)$ of the torus consists of symplectomorphisms that are not Hamiltonian since dy is not exact. Its image under the homomorphism to $H^1(M, \mathbb{R})/\Gamma_\omega$ is the cohomology class of 1-forms, $t[dy]$. As we shall see Hamiltonian torus actions always have fixed points[2].

5.2.2 Fixed Points of Symplectomorphisms

Let M be an n-dimensional real manifold. As we know, the cotangent bundle, $T^\star(M)$, always has a canonical symplectic form, ω. Let $f : M \times M \to \mathbb{R}$ be a smooth function generating a symplectomorphism $\varphi : M \to M$ defined by $\varphi(p, d_p f) = (q, -d_q f)$. We want to describe the fixed points of φ. To do so we introduce the function $\psi : M \to \mathbb{R}$ given by $\psi(p) = f(p, p)$.

Proposition 5.2.13. *There is a $1 : 1$ correspondence between the fixed points of the symplectomorphism φ and the critical points of ψ.*

Proof. At $p_0 \in M$, $d_{p_0}\psi = (d_p f + d_q f)|_{(p,q)=(p_0,p_0)}$. Let $X = d_p f|_{(p,q)=(p_0,p_0)}$. Now, if p_0 is a critical point of ψ, $d_{p_0}\psi = 0$, which happens if and only if $d_q f|_{(p,q)=(p_0,q_0)} = -X$, or equivalently if and only if the point in the graph of φ corresponding to $(p, q) = (p_0, p_0)$ is (p_0, p_0, X, X). In other words $\varphi(p_0, X) = (p_0, X)$ so (p_0, X) is a fixed point. Since all steps are reversible the converse also holds. $\square$

Remark 5.2.14. This situation is somewhat analogous to problems addressed by Newton's method. Here one looks for zeros of a smooth real valued function Ψ defined on an interval $I \subseteq \mathbb{R}$, where Ψ' is never zero. Let

$$\Phi(x) = \frac{x\Psi'(x) - \Psi(x)}{\Psi'(x)}.$$

Then a point $x \in I$ is a zero of Ψ if and only if it is a fixed point of Φ.

---

[2]Conversely, it is a classical result of Frankel that a symplectic S^1-action on a compact Kähler manifold M with at least one fixed point must be Hamiltonian.

We say that $f : M \to M$ is sufficiently C^1 *close* to the identity I, i.e., if it is in some sufficiently small neighborhood U of the identity map in the C^1 topology. Now, $f \in \text{Symp}(M, \omega)$ gives a graph $\Gamma_f = \{(p, f(p)), \ p \in M\} \subseteq (M \times M, \ \pi_1^\star \omega - \pi_2^\star \omega)$ which is a Lagrangian submanifold. If f is C^1 close to the identity, then Γ_f is C^1 close to the diagonal $\Delta = \{(p, p), \ p \in M\} \subseteq (M \times M, \ \pi_1^\star \omega - \pi_2^\star \omega)$ (i.e. the graph of the identity map). By Weinstein's tubular neighborhood theorem (see [27], p. 51) a tubular neighborhood of Δ is diffeomorphic to some $U_0 \subset (T^\star(M), \omega_{T^\star(M)})$, and the graph of f gives a section μ, C^1 close to the zero section. Here $\mu \in \Omega^1(M)$ and since its graph is Lagrangian μ is closed.

This will enable us to prove that many symplectomorphisms of compact symplectic manifolds actually have at least two fixed points.

Theorem 5.2.15. *Let (M, ω) be a compact symplectic manifold such that $H^1(M) = 0$. Then any symplectomorphism of M that is sufficiently C^1 close to the identity has at least two fixed points.*

Proof. If f is a symplectomorphism sufficiently C^1 close to the identity, then its graph corresponds to a closed 1-form α on M. As $H^1(M) = 0$, $\alpha = d\beta$ for some $\beta \in C^1(M)$. Since M is compact, β has at least two critical points. A point p for which $\alpha(p) = d\beta(p) = 0$ corresponds to a point in the intersection of the graph of f with the diagonal, that is, is a fixed point of f. $\square$

5.2.3   Arnold's Conjecture

Consider the symplectic 2-torus $T^2 = \mathbb{R}^2/\mathbb{Z}^2$ whose symplectic structure is inherited from $\mathbb{R}^2$, i.e. the standard area form. Let H be a time independent Hamiltonian on T^2 and ϕ be the time 1 periodic map of the associated flow ($\phi(x + 1) = \phi(x)$, a symplectomorphism). It would be nice to be able to count, or at least estimate the number of fixed points of ϕ. Regarding the critical points of H as the equation of motion, then $(\dot{x}, \dot{y}) = J\nabla H(x, y) = 0$. Here there are at least 2 fixed points since T^2 is a compact space and any nonconstant function on it must achieve its maximum and minimum at two distinct points (actually, it is known there are 3 critical points here). V. Arnold conjectured that similar estimates on the number of fixed points should hold for any compact symplectic manifold. This can be regarded as a generalization of the Poincaré-Birkhoff theorem (also known as Poincaré's *last geometric theorem*) which we discuss in the next section.

Arnold's Conjecture: Let (M, ω) be a compact symplectic manifold and $f : M \to M$ be a Hamiltonian diffeomorphism with nondegenerate fixed points. Then

$$\# \operatorname{Fix}(f) \geq \dim_{\mathbb{R}} H^{\star}(M, \mathbb{R}) = \sum_k \dim_{\mathbb{R}} H^k(M, \mathbb{R}).$$

As we saw in the case of a torus T^2, this statement is false for non-Hamiltonian vector fields. Arnold's conjecture has been proved for tori and a number of other compact symplectic manifolds.

An area preserving smooth transformation of the torus T^2 fixing the center of mass has at least 3, and generically 4, fixed points. In general, for a symplectic transformation of T^{2n}, fixing the center of mass, these numbers are $2n + 1$ and $4n$, respectively. This is the celebrated Conley-Zehnder theorem (see [28]).

Moreover, in his 1984 dissertation B. Fortune proved Arnold's conjecture for complex projective spaces based on some previously unpublished work of Eliashberg (see [38]).

5.3   Poincaré's Last Geometric Theorem

Shortly before his death, Poincaré submitted a paper[3] to Rendiconti del Circolo Mathematica di Palermo [86]. There he writes:

> "...I have never made public a work that is so unfinished; so I believe it is necessary to explain in a few words the reasons that have induced me to publish it and to start with the reasons that brought me to undertake this. Already a long time ago, I have shown the existence of periodic solutions

---

[3]The paper was accepted 3 days later.

of the three-body problem. However, the result is not quite satisfactory, for if the existence of each type of solution had been established for small values of the masses, one did not see what would happen for much larger values, which of the solutions would persist and in which order they would vanish. Thinking about this question I became convinced that the answer would depend on a certain geometric theorem being correct or false, a theorem of which the formulation is very simple, at least in the case of the restricted problem and of dynamics problems that have not more than two degrees-of-freedom..."

Poincaré adds that for two years he had tried to prove this without success. However, he was absolutely convinced that it was correct.

"...It seems that under these conditions, I would have to abstain of all publication of which I had not solved the problems. After all my fruitless efforts of long months, it seemed to me the wisest road to let the problem ripen and put it off my mind for a few years. That would have been very good if I had been certain that I could retake it some time, but at my age I could not say so. Also the importance of the matter is too great and the amount of results obtained already too considerable..."

Although only 58 years old, he had been suffering from serious prostate problems for several years and it seemed to him unwise to keep all these ideas to himself. As it turned out he was right as he died four months later. His last geometric theorem can be considered the first theorem of Symplectic Geometry and is the following:

Theorem 5.3.1. (Poincaré-Birkhoff Theorem). *Consider in $\mathbb{R}^2$ the ring A bounded by the smooth closed curves C_a and C_b (circles). The map $F : A \to A$ is continuous, $1 : 1$ and area preserving. Applying F to A the points of C_a move in the negative sense and those of C_b move in the positive sense. Then F has at least two fixed points.*

This was first proved by George Birkhoff in the same year in [12]. The conditions that F both preserves area, and is a twist map are crucial because a rotation of the annulus is evidently area preserving, but has no fixed points because it does not satisfy the twist condition. On the other hand, the map $(x, y) \mapsto (x^2, x + y - \pi)$ is evidently a twist map without fixed points. The Jacobian is $\neq 1$ so it does not preserve area. We formulate the result as Theorem 5.3.5 and our proof follows Le Calvez and Wang [66].

Let $A = S^1 \times [0, 1]$ be an annulus and $F : A \to A$ be a homeomorphism homotopic to the identity. Let $\widetilde{A} = \mathbb{R} \times [0, 1]$ be the universal cover of A, and $\widetilde{F}$ a lift of F. In addition, we suppose that F satisfies the *boundary twist condition* :

$$\pi_1(\widetilde{F}(x, 0)) < x < \pi_1(\widetilde{F}(x, 1)), \ \forall \, x \in \mathbb{R},$$

where $\pi_1 : \widetilde{A} \to \mathbb{R}$ is the projection map on the first factor and $\mathrm{Fix}_*(F)$ denotes the set of fixed points of F *that are lifted to fixed points of* $\widetilde{F}$.

Let $\gamma : [0, 1] \to M$ be a continuous path in the topological space M. If N_1 and N_2 are two subsets of M, we say that γ joins N_1 to N_2 if $\gamma(0) \in N_1$ and $\gamma(1) \in N_2$. Now, let $\varphi : M \to M$ be a homeomorphism.

Definition 5.3.2. We call γ a *positive path* of φ when

$$\varphi(\gamma(t_2)) \neq \gamma(t_1), \ \text{for any } t_2 \geq t_1.$$

We note that a positive path γ does not meet the fixed point set. Also for any k, the images $\varphi^k \circ \gamma$, are also positive. Now, if γ is a path in the universal cover $\widetilde{A} - \mathrm{Fix}(\widetilde{F})$, we define the *variation of angle* of $\widetilde{F}$ as

$$\iota_\gamma(\widetilde{F}) = \int_{\widetilde{X} \circ \gamma} d\theta,$$

where the vector field, $\widetilde{X} : z \mapsto \widetilde{F}(z) - z$ is invariant under the covering automorphism $T : (x, y) \mapsto (x + 1, y)$ and lifts a vector field X on A whose singular set is exactly $\mathrm{Fix}_*(F)$, and

$$d\theta = \frac{x\,dy - y\,dx}{2\pi(x^2 + y^2)}$$

is the usual polar form on $\mathbb{R}^2 - \{0\}$. The form $d\theta$ being closed can be integrated on any (even nonsmooth) path in $\mathbb{R}^2 - \{0\}$.

Lemma 5.3.3. *If γ is a positive path of $\widetilde{F}$ that joins a boundary line of $\widetilde{A}$ to the other boundary, then,*

$$\iota_\gamma(\widetilde{F}) = -\frac{1}{2}.$$

Proof. We write the proof in the case where γ joins $\mathbb{R} \times \{0\}$ to $\mathbb{R} \times \{1\}$, the other case being similar. The boundary of the simplex

$$\Delta = \{(t_1, t_2) \mid t_1 \leq t_2\},$$

can be written as

$$\partial\Delta = \delta_d \delta_h \delta_v$$

where δ_d is the diagonal, δ_h the horizontal and δ_v the vertical segment.

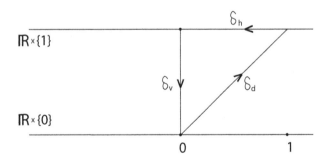

Since γ is a positive path, the map

$$\Phi : (t_1, t_2) \mapsto \widetilde{F}(\gamma(t_2)) - \gamma(t_1)$$

does not vanish on Δ, thus

$$\int_{\Phi\circ\delta_d} d\theta + \int_{\Phi\circ(\delta_h\delta_v)} d\theta = \int_{\Phi\circ\partial\Delta} d\theta = 0.$$

The image by Φ of each segment δ_h and δ_v does not intersect the vertical half-line $\{0\} \times (-\infty, 0]$. Hence

$$\iota_\gamma f = \int_{\Phi \circ \delta_d} d\theta = -\int_{\Phi \circ (\delta_h \delta_v)} d\theta = -\frac{1}{2}.$$

$\square$

Lemma 5.3.4. *Let M be a compact connected (and therefore path-connected) manifold and φ be a homeomorphism without fixed points and with no wandering point. If $N \subset M$ satisfies $\varphi(N) \subset N$, then for every $p \in M$ there exists a positive path of φ that joins N to p.*

Proof. One must prove the equality $M = K$, where K is the set of points that can be joined by a positive path of φ whose origin belongs to N. The space M being connected, it is sufficient to prove that $\overline{K} \subset \mathrm{Int}(K)$. Fix $x \in \overline{K}$. By hypothesis, one can find a path-connected neighborhood U of x such that $\overline{U} \cap \varphi(\overline{U}) = \emptyset$. We will prove $U \subset K$.

Since $x \in K$ there exists a positive path $\gamma_0 : [0,1] \to M$ from N to U. The closures of the subsets $I_1 = \gamma_0^{-1}(U)$ and $I_2 = \gamma_0^{-1}(\varphi(U))$ do not intersect because $\overline{U} \cap \varphi(\overline{U}) = \emptyset$. This implies that $\inf I_1 \neq \inf I_2$. Suppose first that $\inf I_1 < \inf I_2$ (this includes the case where $I_2 = \emptyset$). In that case, there is a subpath γ_1 of γ_0 from N to U that does not meet $\varphi(U)$. For every $p \in U$ one can find a path γ inside U that joins the extremity x_1 of γ_1 to p. The path $\gamma_2 = \gamma_1 \gamma$ is positive because γ_1 is positive and $\varphi(\gamma)$ is disjoint both from γ and γ_1. This implies that $p \in K$.

Suppose now that $\inf I_2 < \inf I_1$. In that case, there is a subpath γ_1 of γ_0 from N to $\varphi(U)$ that does not meet U. We denote by x_1 its extremity. The point $\varphi(x_1)$ does not belong to γ_1 because this path is positive. Since M is a Hausdorff space, the path is compact. Thus we can find a path-connected neighborhood $W \subset \varphi(U)$ of x_1 such that $\varphi(W)$ does not intersect γ_1. Since W is a nonwandering set, we can find a point x_2 in W whose positive orbit meets $\varphi^{-1}(W) \subset U$. Now, choose a path γ inside W that joins x_1 to x_2. The path $\gamma_2 = \gamma_1 \gamma$ does not meet U and is positive because γ_1 is positive and $\varphi(\gamma)$ is disjoint from γ_1 and γ. Consider the integer $m \geq 1$ such that $\varphi^m(\gamma_2) \cap U \neq \emptyset$ and

$\varphi^j(\gamma_2) \cap U = \emptyset$ if $0 \le j < m$. Since by assumption $\varphi(N) \subset N$ the path $\varphi^m(\gamma_2)$ is a positive path from N to U that does not meet $\varphi(U)$. We conclude as in the first case.   □

Now we state and prove the Poincaré-Birkhoff Theorem.

Theorem 5.3.5. *(Poincaré-Birkhoff Theorem.) If F preserves the measure (area) induced by $dx \wedge dy$, $|\mathrm{Fix}_\star(F)| \ge 2$.*

Proof. As above we consider the vector field $\widetilde{X} : z \mapsto \widetilde{F}(z) - z$ invariant under the covering automorphism $T : (x, y) \mapsto (x + 1, y)$ which lifts the vector field X on A whose singular set is exactly $\mathrm{Fix}_\star(F)$.

The theorem will be proved if we find a loop Γ in $\widetilde{A} - \mathrm{Fix}(\widetilde{F})$ such that $\iota_\Gamma \ne 0$. Indeed, if $\mathrm{Fix}_\star(F)$ is finite, i.e. if $\mathrm{Fix}(\widetilde{F})$ is discrete, then

$$\iota_\Gamma(\widetilde{F}) = \sum_{z \in \mathrm{Fix}(\widetilde{F})} i(\widetilde{X}, z) \int_{\xi_z \circ \Gamma} d\theta,$$

where $\iota(\widetilde{X}, z)$ denotes the Poincaré Hopf index of the vector field $\widetilde{X}$ at z and ξ_z is the vector field $z_1 \mapsto z_1 - z$. This implies that $|\mathrm{Fix}_\star(F)| \ge 2$ because $\iota(\widetilde{X}, z) = \iota(X, \pi(z))$ and because the Poincaré-Hopf formula 3.4.11 asserts that

$$\sum_{z \in \mathrm{Fix}_\star(F)} \iota(X, z) = \chi(A) = 0.$$

If we can find two paths γ and γ_1 such that $\iota_\gamma(\widetilde{F}) = \iota_{\gamma_1}(\widetilde{F}) \ne 0$, the first one joining $\mathbb{R} \times \{0\}$ to $\mathbb{R} \times \{1\}$, the second one joining $\mathbb{R} \times \{1\}$ to $\mathbb{R} \times \{0\}$, then the loop $\Gamma = \gamma \delta \gamma_1 \delta_1$ obtained by adding horizontal segments on each boundary line will satisfy

$$\iota_\Gamma(\widetilde{F}) = 2\iota_\gamma(\widetilde{F}) \ne 0.$$

This is what we will do. First, we remark that we can suppose that $\mathrm{Fix}(F)$ does not separate the two boundary circles otherwise we have $|\mathrm{Fix}(F)| = \infty$. If n is large enough, the homeomorphism F' of the annulus $A' = \mathbb{R}/n\mathbb{Z} \times [0, 1]$ lifted by $\widetilde{F}$ has no fixed points, except for the ones that are lifted to fixed points of $\widetilde{F}$. Therefore $\mathrm{Fix}(F')$ = $\mathrm{Fix}_*(F')$ does not separate the boundary circles of A', and one may consider the connected component B of $A' - \mathrm{Fix}(F')$ that contains the boundary. Moreover F' has no wandering point because it preserves area (Theorem 5.1.1). Now, applying Lemma 5.3.4 to $M = B$, $\varphi = F'|_B$ and $N = \mathbb{R}/n\mathbb{Z} \times \{0\}$ or $N = \mathbb{R}/n\mathbb{Z} \times \{1\}$, we obtain a positive path of F' from one of the boundary circles of A' to the other. Such a path lifts to a positive path of $\widetilde{F}$ from the corresponding boundary line to the other. Then, the Poincaré-Birkhoff Theorem follows from Lemma 5.3.3. $\square$

Another fixed point theorem spiritually close to Theorem 5.3.5 is the following: Consider the unit sphere S^2. Since $\chi(S^2) = 2 \neq 0$, by Lefschetz' theorem any $f : S^2 \to S^2$ homotopic to the identity must have at least one fixed point. It can have only 1 as for example in the case of Riemann sphere (see figure), where the translation $z \mapsto z + 1$ has only the North Pole N as fixed point.

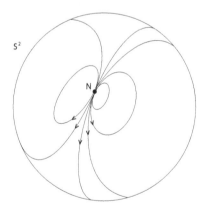

But if f is, in addition, an area preserving map, then it must have at least two. Indeed,

Theorem 5.3.6. *A homeomorphism f of S^2 homotopic to the identity preserving a regular measure μ has at least 2 fixed points. In particular, every diffeomorphism of S^2 leaving an area form ω invariant $(f^\star\omega = \omega)$ has at least 2 fixed points.*

Proof. Let $p \in S^2$ be the fixed point the existence of which is guaranteed as above and $g := f|S^2-\{p\}$. We can identify $\mathbb{R}^2$ with $S^2-\{p\}$ and move the measure μ from the latter to the former. Then the transplanted g also preserves the transplanted measure (which we also call μ).

If g has no fixed point, then by Brouwer's theorem there is an open set U with the property

$$g^i(U) \cap g^k(U) = \emptyset$$

for any $i \neq k$. Arguing as in Poincaré recurrence,

$$\sum_j = (-1)^n \mu(f^j(U)) = n\mu(U) \leq \mu(S^2)$$

for any $n \geq 1$. Therefore $\mu(U) = 0$, a contradiction.

Now, if $f : S^2 \to S^2$ is such that $f^\star \omega = \omega$, the commutativity of the diagram defining the degree shows that f has degree 1 and so is homotopic to the identity (see Section 3.2.2, parts 3, 4). Hence f has at least two fixed points.   □

We also mention a result, which depends on the Poincaré-Birkhoff theorem. Namely, the problem of billiards. A mathematical billiard system consists of a domain in the plane (the billiard table) and a point-mass (a billiard ball) that moves inside the domain along straight lines with constant speed until it hits the boundary. The reflection off the boundary is elastic and subject to the familiar law: the angle of incidence equals the angle of reflection. After the reflection, the point continues in a straight line with its new velocity until it hits the boundary again, etc.

Corollary 5.3.7. *For a billiard system on a smooth convex table, there are at least 2 periodic orbits of type $0 < p/q < 1$ (meaning that T^q winds around the table p times).*

Proof. The map T^q leaves one boundary of the annulus $X = T^1 \times [-1,1]$ fixed, the other boundary is turned around q times. Now define $S(x,y) = (x-1,y)$ which rotates every point around once. Now, $T^q S^{-p}$ rotates one side of the boundary by $-2\pi p$ and the other side of the boundary by $2\pi(q-p)$. Since the boundary is now turned in a different direction, there are fixed points of $T^q S^{-p}$. For such a fixed point $T^q(x,y) = S^p(x,y)$, which is what we called an orbit of type $0 < p/q < 1$. $\square$

5.4 Automorphisms of Lie Algebras

Here we first prove two results concerning automorphisms of finite dimensional real or complex Lie algebras. The first of these due to A. Borel and J. P. Serre in [14], involves fixed points, while the second, due to N. Jacobson in [61], leads to hyperbolic automorphisms which are of importance in global analysis, bringing nilmanifolds and infra-nilmanifolds into the discussion of Anosov diffeomorphisms of (not necessarily toral) manifolds.

Theorem 5.4.1. *Let α be an automorphism finite dimensional real or complex Lie algebra, $\mathfrak{g}$. If α is of prime order and has only 0 as a fixed point, then $\mathfrak{g}$ is nilpotent.*

Theorem 5.4.2. *Let α be an automorphism finite dimensional real or complex Lie algebra, $\mathfrak{g}$. If none of the eigenvalues of α are roots of unity, then $\mathfrak{g}$ is nilpotent.*

We shall deal with both simultaneously.

Proof. In the real case we can just complexify $\mathfrak{g}$ so that we can assume we are dealing with a complex Lie algebra. We can therefore apply a generalization of the third Jordan form to automorphisms of $\mathfrak{g}^4$ as follows (see [52]): $\alpha = \psi \circ \mathrm{Exp}(D)$, where,

1. ψ is an automorphism and is diagonalizable.

---

[4]Because $\mathrm{Aut}(\mathfrak{g})$ is a linear algebraic group this is actually a special case of the following more general result: Any complex linear algebraic group H has the property that each $h \in H$ can be written (uniquely) as $h = du$, where d and $u \in H$, d is diagonalizable, u is unipotent and d and u commute. In fact d and u commute with anything in H which commutes with h.

2. D is a nilpotent derivation of $\mathfrak{g}$.

3. D and ψ commute.

Here we use Exp since this applies to linear operators.

Thus $\mathfrak{g} = \sum V_{\lambda_i}$, the sum of the eigenspaces of α. One sees easily that these eigenspaces are invariant under ψ and Exp(D) and since Exp(D) is unipotent, the eigenvalues of ψ and α coincide. Thus in the second statement we can replace α by ψ and assume α is diagonalizable.

We will show we can do the same in the first statement. Since Exp(D) is unipotent and therefore of infinite order and D and ψ commute, ψ also has prime order. Suppose ψ had a fixed point $v \neq 0$ i.e. $\psi(v) = v$. Since D is nilpotent, $D^k(v) = 0$ for some positive integer k. Let k be the least such integer. Applying α we see $D^k(\alpha(v)) = 0$. If $D^j(\alpha(v)) = 0$ for an exponent $j \geq k$, applying α^{-1} (which commutes with α and hence also its Jordan components) to this tells us $D^j(\alpha(v)) = 0$ and so $j = k$. Thus the same exponent works for both v and $\alpha(v)$. Applying D^k we see $D^k\psi(v) = D^k(v) = 0 = D^{k-1}\alpha(v)$, which implies $\alpha(v) = 0$ and so α would have a nonzero fixed point, a contradiction. Thus ψ also has prime order and only zero as a fixed point.

Thus in both Theorems 5.4.1 and 5.4.2, we can and do assume α is diagonalizable. Let V_λ and V_μ be two of the distinct (geometric) eigenspaces. First we observe that in case two it is impossible for $\lambda\mu^j = \lambda\mu^k$, where j and k are positive integers unless $j = k$. For suppose this occurred. Since α is an automorphism and therefore invertible and λ is an eigenvalue it is nonzero. Hence $\mu^j = \mu^k$. Assuming $j \geq k$, $\mu^{j-k} = 1$. So, in case two, since no eigenvalue is a root of unity, $j = k$. In particular, there are infinitely many of these $\lambda\mu^j$.

In case one we also show $\{\lambda\mu^j : j \geq 1\}$ is infinite. Let $v_\lambda \in V_\lambda$ and $v_\mu \in V_\mu$. Then $\alpha[v_\lambda, v_\mu] = \lambda\mu[v_\lambda, v_\mu]$. It follows that $[V_\lambda, V_\mu] = (0)$, or $\lambda\mu$ is an eigenvalue of α. Let $a \in V_\mu$ be fixed. By induction on k either

$$\mathrm{ad}_a^k(v_\lambda) = \mathrm{ad}_a(\mathrm{ad}_a^{k-1})(v_\lambda) = 0,$$

or each $\lambda\mu^k$ is an eigenvalue of α. But since this set is infinite and α has only finitely many eigenvalues, there must be a k large enough so that

$\mathrm{ad}_a^k(v_\lambda) = (0)$. Note that $\mathrm{ad}_a^k(v_\mu)$ is also (0). Let k_0 be the maximum of this finite set of k over all the eigenspaces V_λ. Since $\mathfrak{g} = \sum V_\lambda$ we get $\mathrm{ad}_a^{k_0}(\mathfrak{g}) = (0)$ and since $a \in V_\mu$ was arbitrary, $\mathrm{ad}_{V_\mu}^{k_0}(\mathfrak{g}) = (0)$. But μ is an arbitrary eigenvalue of α so the decomposition $\mathfrak{g} = \sum V_\mu$ yields, $\mathrm{ad}_{\mathfrak{g}}^{k_0} = (0)$ which means $\mathfrak{g}$ is nilpotent. □

In the special case of Theorem 5.4.2 when all the eigenvalues λ of α satisfy $|\lambda| > 1$, one gets a useful estimate on the index of nilpotence of $\mathfrak{g}$. Namely,

$$\mathrm{Ind\,Nilp}(\mathfrak{g}) \leq \frac{\log |\lambda_+|}{\log |\lambda_-|},$$

where $|\lambda_+|$ and $|\lambda_-|$ are respectively the largest and smallest of the moduli of the eigenvalues of α (see Moskowitz [72]).

This bound is sharp as it is achieved in the Lie algebra of full nil triangular matrices (over $\mathbb{R}$, or $\mathbb{C}$). Of course, if $\mathfrak{g}$ had an automorphism all of whose eigenvalues satisfied $|\lambda| < 1$, then its inverse would be an automorphism all of whose eigenvalues have modulus greater than 1 and so $\mathfrak{g}$ would be nilpotent. Finally, it follows easily from this that a connected and simply connected 2-step nilpotent Lie group G always possesses a *contracting family* of automorphisms. Namely, a multiplicative 1-parameter group $t \mapsto \alpha_t$ satisfying $\lim_{t\to 0} \alpha_t(g) = 1$, for each $g \in G$.

Let $\mathfrak{g}$ be a real or complex Lie algebra and $\mathrm{Der}\,\mathfrak{g}$ be its algebra of derivations.

Corollary 5.4.3. *Let $\mathfrak{g}$ be a real or complex Lie algebra and D be a derivation none of whose eigenvalues is pure imaginary. Then $\mathfrak{g}$ is nilpotent.*

This is because $\mathrm{Exp}(D)$ is an automorphism $\mathfrak{g}$ which has no eigenvalues of absolute value 1.

Definition 5.4.4. We say a group H is *quasi cyclic* if it has a finite sequence of normal subgroups

$$(1) \subseteq H_1 \subseteq \ldots \subseteq H_n,$$

whose successive quotients H_i/H_{i-1} are either finite cyclic or 1 dimensional tori. For example, any finite abelian group is quasi cyclic.

Theorem 5.4.1 was used to prove the following,

Corollary 5.4.5. *Let H be a quasi cyclic subgroup of a compact connected Lie group G and T be a maximal torus of G. Then H normalizes T.*

We now turn to a fixed point theorem concerning simply connected nilpotent Lie groups N with Lie algebra $\mathfrak{n}$ and center $Z(N)$. Let $\mathrm{Aut}(N)$ denote the smooth automorphisms of N. Important facts about such groups (see [1]) are exp : $\mathfrak{n} \to N$ is a global diffeomorphism and $\mathrm{Aut}(N)$ is equivariantly equivalent to $\mathrm{Aut}(\mathfrak{n})$ with respect to exp. Also, $Z(N)$ is a closed, connected, characteristic (stable under $\mathrm{Aut}(N)$) subgroup of N.

The following result can be regarded as a fixed point theorem (It is a fixed point theorem for the function $L_n \circ \alpha$.) and plays a role in the study of expanding affine maps of nilmanifolds. The quantity on the right side of the equation below is called the *displacement* of n_0 by α. For this statement to be valid it is essential that the simply connected group be nilpotent. One checks rather easily that it fails for the $ax + b$ group.

Theorem 5.4.6. *For a simply connected nilpotent Lie group N and $\alpha \in \mathrm{Aut}(N)$ with 1 not an eigenvalue[5] of $\alpha_\star$, there is a unique $n_0 \in N$ so that $n = \alpha(n_0)n_0^{-1}$. Moreover, 1 is the only fixed point of α.*

Proof. For the second statement, if $\alpha(g) = g$, then taking the derivative $\alpha_\star(X) = X$, where $X = \log g$. By hypothesis X must be 0. Hence $g = 1$. From this, uniqueness of n_0 follows. For if $\alpha(g)g^{-1} = n = \alpha(h)h^{-1}$, then $\alpha(h^{-1}g) = h^{-1}g$. As a fixed point of α, $h^{-1}g = 1$ so $g = h$. We now turn to existence.

As above, another way to say this is, for a given $n \in N$ there is an $n_0 \in N$ so that $n\alpha(n_0) = n_0$. We prove this by induction on the index

---

[5]In other words α is of Lefschetz type.

of nilpotence of N. When this index is 1, that is, when N is abelian, we seek an n_0 so that $n + \alpha(n_0) = n_0$. Thus we want to find a (unique) solution to the equation $(\alpha - I)(n_0) = -n$. Passing to the Lie algebra, this is $(\alpha_* - I)(X_0) = -Y$, where $X_0 = \log(n_0)$ and $Y = \log n$. Since 1 is not an eigenvalue of α_*, $\alpha_* - I$ is invertible and so we can solve $(\alpha_* - I)(X_0) = -Y$ uniquely; $X_0 = -(\alpha_* - I)^{-1}(Y)$.

Now suppose the result is true for all simply connected nilpotent groups of lower index of nilpotence. Since $Z(N)$ is connected and non-trivial $N/Z(N)$ is such a group. Moreover α preserves $Z(N)$. Applying the projection $\pi : N \to N/Z(N)$ to the equation $n\alpha(n_0) = n_0$ yields $\bar{n}\bar{\alpha}(\bar{n}_0) = \bar{n}_0$. Since $\bar{\alpha}_*$ also doesn't have 1 as an eigenvalue, by induction we can solve this equation for $\bar{n}_0 \in N/Z(N)$. As π is onto, let n_0 be anything mapping to $\bar{n}_0$. Then $n\alpha(n_0) = zn_0$ for some $z \in Z(N)$ (where the z depends on which preimage one takes). Because $\alpha|Z$ also doesn't have 1 as an eigenvalue, by the abelian case for a given z there is a $z_0 \in Z(N)$ with $z = z_0\alpha(z_0^{-1})$. Thus, $n\alpha(n_0) = z_0\alpha(z_0^{-1})n_0$, which means $n\alpha(z_0n_0) = z_0n_0$. $\square$

5.5   Hyperbolic Automorphisms of a Manifold

Let M be a smooth compact manifold and we will consider $\mathrm{Diff}(M)$ equipped with the topology of uniform C^r convergence (meaning uniform convergence of functions and all their derivatives up to order $r \le \infty$). An *endomorphism* is a smooth map $f : M \to M$. Such an f is called an *automorphism* if $f \in \mathrm{Diff}(M)$. The map

$$f^k := f \circ f \circ \cdots \circ f : M \to M$$

is called the kth *iterate* of f. A point x in M is called a *periodic point* relatively to f, if there exists an $k \in \mathbb{Z}^+$ so that $f^k(x) = x$. The smallest positive integer k satisfying this is called the *minimal period* of the point x. If every point in M is a periodic point with the same minimal period k, then we say that f is a *periodic map* with minimal period k. When $k = 1$ this is just a *fixed point*. We denote the set of fixed points (resp. periodic points) of f by $\mathrm{Fix}(f)$ (resp. $\mathrm{Per}(f)$) and by $N_n(f)$ the number of points of minimal period n. As observed earlier, since M is compact and f is a smooth automorphism, its regular fixed points are isolated and therefore finite in number.

The *orbit* of a point $x \in M$ relative to the endomorphism f is the set $\{x, f(x), ..., f^k(x), ...\}$. If x is a periodic point of f of minimal period k, then its orbit is $\{x, f(x), ..., f^{k-1}(x)\}$, and we will call this a k-*cycle*. Evidently,

$$\text{Fix}(f) \subset \text{Per}(f) \subset \Omega(f) \quad \text{(the nonwandering points).}$$

A linear automorphism $A : V \to V$ of a real finite dimensional vector space V is called *hyperbolic* if its eigenvalues λ_i satisfy $|\lambda_i| \neq 1$ for all i (complex eigenvalues are permitted). The map A will be called *contracting* if $|\lambda_i| < 1$ for all i, and *expanding* if $|\lambda_i| > 1$ for all i (*saddle type* otherwise). We remark that for a hyperbolic $A : V \to V$ (which extends to the complexification) we have a canonical A-invariant splitting of $V^{\mathbb{C}}$:

$$V^{\mathbb{C}} = V^s \oplus V^u,$$

where V^s is the eigenspace of A corresponding to eigenvalues of absolute value less than 1 and V^u the eigenspace of those with absolute value greater than 1. A restricted to V^s is contracting and restricted to V^u is expanding.

Now suppose $f : M \to M$ is a diffeomorphism with a fixed point $p \in M$. The derivative $df(p) : T_p(M) \to T_p(M)$ is a linear automorphism of the tangent space of M at p and we say p is a *hyperbolic fixed point* of f, or simply a hyperbolic fixed point, if $df(p)$ is hyperbolic in the sense above.

We will call a periodic point p of period k of f *hyperbolic* if it is a hyperbolic fixed point of f^k. Similarly, p is a *contracting*, or *expanding* periodic point if $df^k(p)$ is a contracting (or expanding) linear automorphism.

An important notion in dynamics is the following:

Definition 5.5.1. Let M be a manifold and f and $g \in \text{Diff}(M)$. We say f and g are *topologically conjugate* if there exists a homeomorphism φ of M such that $g = \varphi \circ f \circ \varphi^{-1}$. Such a homeomorphism φ is called a *topological conjugacy* between f and g.

If φ is a conjugacy between f and g, then it is also a conjugacy between f^k and g^k for $k \in \mathbb{Z}^+$:

$$g^n = \varphi \circ f^n \circ \varphi^{-1}.$$

In particular, φ maps periodic points of f to periodic points of g.

A simple example of such a conjugacy is given by $M = \mathbb{R}$, $f(x) = 2x$ and $g(x) = 8x$. One checks easily that $\varphi(x) = x^3$ is a topological conjugacy. Notice that φ is merely a homeomorphism and not a diffeomorphism. Less trivial examples of topological conjugacy are given in a theorem of Franks concerning Anosov diffeomorphisms of a torus $\mathbb{T}^n$ as well as by a theorem of Manning in the case of a compact nil manifold G/Γ (Theorem 5.5.14), both below.

5.5.1   The Case of a Torus

We now consider the case when M is the n-torus $\mathbb{T}^n$ which can be regarded as the abelian group $\mathbb{R}^n/\mathbb{Z}^n$. Let A be an $n \times n$ matrix with integer entries. Since it leaves $\mathbb{Z}^n$ invariant, it induces an endomorphism

$$\widetilde{A} : \mathbb{T}^n \to \mathbb{T}^n$$

defined by

$$x \mapsto \widetilde{A}(x) := Ax + \mathbb{Z}^n.$$

Definition 5.5.2. If A is an integer $n \times n$ matrix, we call the map $\widetilde{A}$ a *toral endomorphism*. In addition, if $\det(A) = \pm 1$ we call $\widetilde{A}$ a *toral automorphism*.

We now calculate the group of all toral automorphisms,

Proposition 5.5.3. $\mathrm{Aut}(\mathbb{T}^n) = \mathrm{SL}^{\pm}(n, \mathbb{Z})$.

Proof. Consider the natural projection map $\pi : \mathbb{R}^n \to \mathbb{T}^n$. This is the universal covering of $\mathbb{T}^n$ since $\mathbb{R}^n$ is simply connected. Hence each $\alpha \in \mathrm{Aut}(\mathbb{T}^n)$ can be lifted to $\widetilde{\alpha}$ so that the following diagram is commutative.

$$
\begin{array}{ccc}
\mathbb{R}^n & \xrightarrow{\ \widetilde{\alpha}\ } & \mathbb{R}^n \\
\downarrow{\scriptstyle \pi} & & \downarrow{\scriptstyle \pi} \\
\mathbb{T}^n & \xrightarrow{\ \alpha\ } & \mathbb{T}^n
\end{array}
$$

Obviously, $\widetilde{\alpha}$ is an automorphism of $\mathbb{R}^n$ so that it lies in $\mathrm{GL}(n, \mathbb{R})$. But not all automorphisms of $\mathbb{R}^n$ come in this way. For this to happen $\widetilde{\alpha}$ and its inverse must leave $\mathbb{Z}^n$-invariant. That is, in the standard basis, the matrix for $\widetilde{\alpha}$ and its inverse have integers entries and $\det(\widetilde{\alpha})$ is a unit of $\mathbb{Z}$; that is, ± 1. $\qquad\Box$

$\widetilde{A}$ is a *hyperbolic toral automorphism* if A has no eigenvalue on the unit circle. Note that A may have eigenvalues on the unit circle other than roots of unity.

Theorem 5.5.4. *If $\widetilde{A} : \mathbb{T}^n \to \mathbb{T}^n$ is a hyperbolic toral automorphism, then its periodic points are exactly those with rational coordinates.*

Proof. Suppose that $[x] = [(x_1, ..., x_n)] \in \mathbb{T}^n$ is a periodic point, i.e. there is some k such that $\widetilde{A}^k([x]) = [x]$ which means $\widetilde{A}^k([x]) = x + \mathbb{Z}^n = I \cdot x + \mathbb{Z}^n \Longleftrightarrow (\widetilde{A}^k - I)x = \overrightarrow{m}$, where $\overrightarrow{m} \in \mathbb{Z}^n$. Since $\widetilde{A}$ is hyperbolic, A has no eigenvalues of modulus 1. Hence the same is true of $\widetilde{A}^k$. Thus, there is no vector $v \neq 0$ such that $(\widetilde{A}^k - I)v = 0$, that is $(\widetilde{A}^k - I)$ is invertible. Hence, we can solve the equation $(\widetilde{A}^k - I)x = \overrightarrow{m}$ and get

$$x = (\widetilde{A}^n - I)^{-1} \cdot \overrightarrow{m}.$$

Now, since A has entries in $\mathbb{Z}$, $(\widetilde{A}^k - I)$ has entries in $\mathbb{Z}$, therefore $(\widetilde{A}^k - I)^{-1}$ has entries in $\mathbb{Q}$, i.e. all components of x are rational.

We now prove the converse, i.e. that any point $\mathbb{T}^n$ with rational coordinates is a fixed point for $\widetilde{A}$. For a fixed positive integer k denote by $\mathcal{R}(k)$ the set of rational points in $\mathbb{T}^n$ with denominator k, i.e.

$$\mathcal{R}(k) = \pi\left(\left\{ \left(\frac{j_1}{k}, ..., \frac{j_n}{k}\right) \ : \ j_l \in \mathbb{Z}, \ l = 1, ..., n \right\}\right) \subset \mathbb{T}^n$$

where π is the projection map. Then,

$$\widetilde{A}(\mathcal{R}(k)) \subset \mathcal{R}(k).$$

But the set $\mathcal{R}(k)$ has a finite number of points for each k, and since $\widetilde{A}$ is $1:1$, $\widetilde{A}\,|\mathcal{R}(k)$ is a permutation of $\mathcal{R}(k)$. Hence every point in this set is periodic. $\square$

Finally, since as is well known the set $\bigcup_k \mathcal{R}(k)$ is dense in the torus, it follows that the set of periodic points is dense in $\mathbb{T}^n$.

Let $\widetilde{A}$ be a hyperbolic toral automorphism. We will now compute the number of its periodic points of period $k \geq 1$. Let A be the corresponding $n \times n$ integer matrix and let N_k denote the (finite) number of isolated fixed points $\mathbb{T}^n$ of $\widetilde{A}$. The basic idea in the proof of the following theorem is to check that, when no eigenvalue of A^k is equal to 1, by counting the number of lattice points of $\mathbb{Z}^n$ in a fundamental domain of the lattice $(I - A^k)\mathbb{Z}^n$ we get $N_k = |\det(I - A^k)|$. More precisely we have:

Theorem 5.5.5. *Let A be an $n \times n$ matrix with integer entries with the property that it has no eigenvalues equal to a root of unity. Let $\widetilde{A}$ denote the endomorphism that A induces on the n-torus, $\mathbb{T}^n$. Then for each $k \geq 1$, $N_k = |\det(A^k - \mathrm{I})|$.*

Proof. For $x = (x_1, ..., x_n) \in \mathbb{R}^n$ we write $\bar{x} = x + \mathbb{Z}^n$ and without loss we can consider $0 \leq x_i < 1$. Now $\bar{x}$ is fixed under $\widetilde{A}^k$ if and only if $A^k x = x \bmod(\mathbb{Z}^n)$, or if and only if $(A^k - I)x \in \mathbb{Z}^n$. Since by hypothesis A^k has no eigenvalue equal to 1, that is $\det(A^k - I) \neq 0$, it follows that $A^k - I$ is invertible. Thus there is a $1:1$ correspondence between the points x in the half open unit cube satisfying $(A^k - I)x \in \mathbb{Z}^n$ and the integral lattice points contained in the image of the half open unit cube under $A^k - I$. Considering the volume interpretation of an absolute value of the determinant, this number is exactly $|\det(A^k - I)|$. Indeed, if we consider the restriction $(A^k - I)|\mathbb{Z}^n$, the number of integral lattice points in the image of the half open unit cube under $A^k - I$ is equal to the order of the group $\mathbb{Z}^n/\mathrm{Im}\big((A^k - I)|\mathbb{Z}^n\big)$. Now, if $p = |\det(A^k - I)|$, then p is an integer and for $x \in \mathbb{Z}^n$, $x \in \mathrm{Im}(A^k - I)|\mathbb{Z}^n$ if and only if $x = py$ for some $y \in \mathbb{Z}^n$. $\square$

As a specific example we will find N_k in the case of the 2 torus, $\mathbb{T}^2$.

Proposition 5.5.6. *Let $\widetilde{A} : \mathbb{T}^2 \to \mathbb{T}^2$ be a hyperbolic toral automorphism with associated matrix A with $\det(A) = 1$ and eigenvalues λ_1 and λ_2. Then*

$$N_k = |\lambda_1^k + \lambda_2^k - 2|.$$

Proof. A periodic point (x_1, x_2) of period k satisfies

$$A^k \begin{pmatrix} x_1 \\ x_2 \end{pmatrix} = \begin{pmatrix} x_1 \\ x_2 \end{pmatrix} + \begin{pmatrix} m_1 \\ m_2 \end{pmatrix}, \tag{5.1}$$

where m_1, $m_2 \in \mathbb{Z}$. Now the map $A^k - I$ sends the unit square $[0,1) \times [0,1)$ to the parallelogram P generated by the vectors $(A^k - I)e_1$, and $(A^k - I)e_2$, where $e_1 = (1,0)$ and $e_2 = (0,1)$. The integral solutions of the equation (5.1) correspond to the lattice points in $\mathbb{Z}^2 \cap P$. By Pick's theorem (see Theorem 5.5.7 below), the number of these integral solutions is the area of P; namely $|\det(A^k - I)|$. To calculate it, we find the eigenvalues of $A^k - I$. Let v be such an eigenvector. Then $(A^k - I)v = \lambda v$ so that $A^k v = (\lambda + 1)v$. Thus v is also an eigenvector of A^k with eigenvalue $\lambda + 1$. Since $\det(A) = 1$, $\lambda_2 = \lambda_1^{-1}$. Hence, the eigenvalues of $A^k - I$ are $\lambda_1^k - 1$ and $\lambda_2^k - 1 = \lambda_1^{-k} - 1$. Therefore

$$N_k = |(\lambda_1^k - 1)(\lambda_2^k - 1)| = |\lambda_1^k + \lambda_2^k - 2|.$$

$\square$

Here we used the following result[6]

Theorem 5.5.7. (Pick's Theorem.) *The area of a simple polygon (one which has no holes, and whose edges do not intersect) $P \subset \mathbb{R}^2$ and*

---

[6]First published in 1899 by Georg Alexander Pick, this theorem gives an elegant formula for the area of a simple lattice polygon. Pick, born in 1859 in Vienna, was a professor at the University of Prague, when in 1910, Einstein applied for a professorship in Theoretical Physics. Pick was a member of the appointments committee and was the driving force in getting Einstein accepted. They become close friends, frequently playing the violin together. In 1929, Pick returned to Vienna. After the Anschluss, in an attempt to escape the Nazis, Pick returned to Prague. He was captured on July 13, 1942, transported to the Theresienstadt concentration camp, where he died 13 days later. He was 82.

whose vertices are lattice points of $\mathbb{Z}^2$, *is given by*

$$Area(P) = a + \frac{b}{2} - 1$$

where a is the number of lattice points in the interior of P, and b is the number of lattice points of the polygon's perimeter.

For a proof of this result see Varberg ([100]), or Murty and Thain ([80]).

5.5.2   Anosov Diffeomorphisms

A motivation for this section is the following. Consider again a hyperbolic toral automorphism $f_A := \widetilde{A}$, with $A \in \mathrm{SL}(2,\mathbb{Z})$.
    Then,

$$A := \begin{pmatrix} a & b \\ c & d \end{pmatrix}$$

and the characteristic polynomial is

$$\chi_A(\lambda) = \lambda^2 - \mathrm{Tr}(A)\lambda + \det(A).$$

Thus, the two eigenvalues of A are:

$$\lambda_{1,2} = \frac{\mathrm{Tr}(A) \pm \sqrt{\mathrm{Tr}^2(A) - 4}}{2}.$$

Hence,

1. when $\mathrm{Tr}(A) = 0, 1, -1$ we have complex conjugate roots,

2. when $\mathrm{Tr}(A) = \pm 2$ we have two equal real roots,

3. and when $|\mathrm{Tr}(A)| > 2$ we have distinct positive roots.

In the third case, the matrix A has an eigenvalue greater than 1, which we call the *unstable eigenvalue* and denote by λ_u and another one less than 1, which we call the *stable eigenvalue*, denoted by λ_s. The corresponding eigenvectors v_u and v_s span the linear subspaces E_u

and E_s which we call the *unstable eigen-direction* and the *stable eigen-direction*, respectively.

We observe that here the discriminant cannot be a perfect square, for if it were then we would have a Pythagorean triangle with side 2 which cannot be. Hence both eigenvalues are always irrational numbers and therefore the corresponding eigen-directions E_u and E_s have irrational slopes. This means that $E_s \cap \mathbb{Z}^2 = \{(0)\}$ and similarly for E_u. To see this, suppose $0 \neq w \in E_s \cap \mathbb{Z}^2$. Then $A^n \cdot w = \lambda_s^n w \to 0$ as $n \to \infty$, which contradicts the assumption that A is invertible. A similar argument works for the subspace E_u (using A^{-1}).

In addition we can prove that not only do the two complementary directions E_u and E_s have irrational slopes, but they satisfy a stronger arithmetic property: for any $(m, l) \in \mathbb{Z}^2$, with $m \neq 0$, there is $\varepsilon > 0$ and $N(\varepsilon) > 0$ such that

$$\left| \lambda_{s,u} - \frac{l}{m} \right| \geq \frac{N(\varepsilon)}{|m|^{2+\varepsilon}}.$$

The meaning of this is that in order for the (eigen-directions) lines E_u or E_s to come close to a nonzero point of the lattice $\mathbb{Z}^2$, that point must be far from the origin.

Finally, we point out that a toral automorphism is *symplectic*, which means (see Lemma 5.2.5) that setting

$$J := \begin{pmatrix} 0 & 1 \\ -1 & 0 \end{pmatrix},$$

any A with $\det(A) = 1$ must satisfy

$$A^t J A = J.$$

Whenever we have an automorphism f_A which stretches by a factor λ in one direction and shrinks by the same factor in a complementary direction we say that the map f_A is *Anosov*. This example affords the following generalization.

Definition 5.5.8. An *Anosov diffeomorphism* $f : M \to M$ is a diffeomorphism which satisfies the following conditions:

1. There is a continuous splitting of the (complexified) tangent bundle
$$T(M) = E^s \oplus E^u$$
which is preserved by the derivative df,

2. There exist constants $c_1 > 0$, $c_2 > 0$ and $\lambda \in (0,1)$ and a Riemannian metric $\| \cdot \|$ on $T(M)$ such that
$$\|df^n(v)\| \le c_1 \lambda^n \|v\| \ \text{for} \ v \in E^s$$
and
$$c_2 \lambda^{-n} \|v\| \le \|df^n(v)\| \ \text{for} \ v \in E^u.$$

E^s and E^u are called the *contracting* and the *expanding* subbundles of $T(M)$, respectively, and a splitting of the tangent bundle satisfying the above conditions is called *hyperbolic*.

Being an Anosov diffeomorphism is independent of the choice of the Riemannian metric.

Proposition 5.5.9. *For a compact manifold M and its Riemannian (complexified) tangent bundle $T(M)$, the property of being contracting or expanding for $\phi : T(M) \to T(M)$ is independent of the Riemannian metric on $T(M)$.*

This is because the two norms $\| \cdot \|_1$ and $\| \cdot \|_2$ on fibers of $T(M)$ are related locally and hence globally by,
$$\frac{1}{c} \| \cdot \|_1 \le \| \cdot \|_2 \le c \| \cdot \|_1$$
for some $c > 0$.

It is an open question to determine the class of manifolds which admit an Anosov diffeomorphism, as it is difficult to construct examples of this type of map. The most common example is the hyperbolic toral automorphisms. J. Franks proved that any Anosov diffeomorphism on a torus $\mathbb{T}^n$ is topologically conjugate to such a hyperbolic toral automorphism. More precisely, we mention without proof the following theorem and corollary of Franks ([40]).

Theorem 5.5.10. *If $f : \mathbb{T}^n \to \mathbb{T}^n$ is an Anosov diffeomorphism with $\Omega(f) = \mathbb{T}^n$ and if*

$$f^{\star,1} : H^1(\mathbb{T}^n, \mathbb{R}) \to H^1(\mathbb{T}^n, \mathbb{R}),$$

is hyperbolic, then f is topologically conjugate to a hyperbolic toral automorphism.

Corollary 5.5.11. *If $f : \mathbb{T}^n \to \mathbb{T}^n$ is an Anosov diffeomorphism and*

$$f^{\star,1} : H^1(\mathbb{T}^n, \mathbb{R}) \to H^1(\mathbb{T}^n, \mathbb{R}),$$

is hyperbolic, then f has a fixed point.

Let M be a compact manifold, and $f : M \to M$ a diffeomorphism. From Lefschetz' Theorem (see Definition 3.2.10 and Theorem 3.3.2) we know

$$L(f) = \sum_i (-1)^i \operatorname{Tr}(f^{\star,i}) = \sum_{p \in \operatorname{Fix}(f)} \nu_p,$$

where $\nu_p = \operatorname{sgn}\det(\mathrm{I} - df_p)$. Now, assume that f has only hyperbolic fixed points, then

Proposition 5.5.12. *If $p \in \operatorname{Fix}(f)$ we have*

$$\nu_p = (-1)^u \Delta$$

where u is the dimension of the eigenspace E^u corresponding to eigenvalues μ such that $|\mu| > 1$ and $\Delta = 1$ if df_p preserves orientation on E^u and $\Delta = -1$ if df_p reverses it.

Proof. We have

$$\det(\mathrm{I} - df_p) = \prod_i (1 - \lambda_i) \prod_j (1 - \mu_j) \prod_k (1 - \xi_k)$$

where λ_i, and μ_j are the real eigenvalues of df_p with $|\lambda_i| < 1 < |\mu_j|$ for all i, j, and ξ_k are the complex ones. Now, as $1 - \lambda_i > 0$ for all i and the eigenvalues ξ_k come as conjugate pairs this implies

$$\prod_k (1 - \xi_k) = \prod_k (1 - \xi_k)(1 - \bar{\xi}_k) = \prod_k |1 - \xi_k|^2.$$

Hence,

$$\nu_p = \operatorname{sgn} \prod_j (1 - \mu_j).$$

This shows $\nu_p = (-1)^\tau$ where τ is the number of positive eigenvalues bigger than 1. If σ is the number of negative eigenvalues < -1, we can write $\nu_p = (-1)^{\tau + 2\sigma} = (-1)^u (-1)^\sigma$. Now, $(-1)^\tau = \Delta$ since a negative eigenvalue reverses orientation, and the complex eigenvalues correspond to rotations and thus does not affect the orientation.   □

5.5.3   Nilmanifold Examples of Anosov Diffeomorphisms

Definition 5.5.13. Let G be a connected and simply connected nilpotent Lie group and Γ be a discrete cocompact subgroup. Then we call the compact manifold G/Γ a *nilmanifold*.

For example, the torus $\mathbb{T}^n := \mathbb{R}^n / \mathbb{Z}^n$ is a nilmanifold. Since nilpotent nonabelian Lie groups are of dimension ≥ 3 (see (3.8) of [1]), tori are the only compact nilmanifolds of dimension ≤ 2.

Let M be a compact smooth manifold, $f : M \to M$ a smooth map and suppose M consists of only nonwandering points of f. In [92], p. 762, Smale sets himself the task of finding nontoral examples of Anosov diffeomorphisms of M, and perhaps even finding all such diffeomorphisms of M (all up to topological conjugacy). His construction goes as follows:

Let G be a connected, simply connected Lie group with a uniform lattice Γ and Lie algebra $\mathfrak{g}$ and suppose $f_0 : G \to G$ is a smooth automorphism of G preserving Γ, thus giving rise to a smooth self map f of G/Γ. Identifying $\mathfrak{g}$ with $T_1(G)$ and assuming also that

$$d(f_0)(0) : \mathfrak{g} \to \mathfrak{g},$$

is *hyperbolic*, by Theorem 5.4.2 we see that G must be *nilpotent*. Then using the commutativity of the diagram involving the exponential map of G we see the self map df_Γ of $T_\Gamma(G/\Gamma)$ is hyperbolic. Hence f_0 induces an Anosov diffeomorphism, f, on the compact space G/Γ.

The only question remaining in the above construction is: When does such a map f_0 exist? Since G must be nilpotent, the existence of Γ devolves to well known results of Malcev (see [67]). Namely, a necessary and sufficient condition for a simply connected nilpotent Lie group G to have a uniform lattice is that $\mathfrak{g}$ have a rational form. Moreover if Γ is a uniform lattice in G, then any automorphism of Γ extends uniquely to an automorphism of G. Hence it will be sufficient to look for a hyperbolic nilpotent Lie group automorphism $f : G \to G$ leaving Γ invariant and where $\mathfrak{g}$ has a rational form. One can then apply Manning's theorem [92].

Theorem 5.5.14. *Let $d : G/\Gamma \to G/\Gamma$ be an Anosov diffeomorphism of a nilmanifold, then d is topologically conjugate to a hyperbolic nilmanifold automorphism $f : G/\Gamma \to G/\Gamma$.*

To find more of these examples one needs the concept of an infra-nilmanifold. Dealing with this would take us too far afield and the interested reader is referred to [92]. However, we note that the only known manifolds admitting Anosov diffeomorphisms are nilmanifolds and infra-nilmanifolds leading to the following conjecture of Gorbatsevich (see [46]):

Gorbatsevich Conjecture: If a compact manifold M admits an Anosov diffeomorphism, then M is homeomorphic to a nilmanifold, or an infra-nilmanifold.

We close this subsection by mentioning the following important theorem of Anosov, [4].

Theorem 5.5.15. *(Anosov Theorem) For smooth self map f of a nilmanifold M*

$$N(f) = |L(f)|.$$

We remind the reader that $N(f)$, the *Nielsen number* is defined as follows. We set an equivalence relation in Fix(f). We say that $x \sim y$ if and only if there exists a path γ from x to y such that $f(\gamma)$ is homotopic to γ. The equivalence classes with respect to this relation are called the *Nielsen classes* of f, and the Nielsen number, $N(f)$, is defined as the number of Nielsen classes having nonzero fixed point index sum.

As a special case of this we have the following result (see [20]).

Corollary 5.5.16. *A compact connected Lie group G with the property that $N(f) = |L(f)|$ for all smooth maps $f : G \to G$ is a torus.*

The above result depends on the fact that $\pi_1(G)$ cannot have elements of finite order. Similarly, when $M = G/\Gamma$ is a compact nilmanifold (see above) its fundamental group, Γ is also torsion free.

Could it be that if M is a compact, connected manifold whose fundamental group has no elements of finite order, then for any $f \in \text{Diff}(M)$, $N(f) = |L(f)|$? Or if not, then perhaps this is true when $M = G/\Gamma$, where G is a connected and simply connected Lie group and Γ is a cocompact discrete subgroup.

We remark that most manifolds do not admit an Anosov diffeomorphism. Among these are Homotopy spheres, real, complex and quaternion projective spaces and lens spaces.

5.6 The Lefschetz Zeta Function

For a smooth compact manifold M and a continuous self map $f : M \to M$, let N_k denote the number of isolated fixed points of f^k (permitting f^k to have an infinite number of fixed points), and if $N_k < +\infty$ for any $k = 1, 2, ...$, then, following Artin-Mazur, one defines the *zeta function*[7] of f as the formal power series

$$\zeta_f(z) = \exp\left(\sum_{k=1}^{\infty} \frac{N_k}{k} z^k \right).$$

Artin and Mazur proved in [5] that for a dense set of $\text{Diff}(M)$, the zeta function has a positive radius of convergence, so it can really be considered a function.

This function turns out to be an important invariant of f. Evidently, $\zeta_f(z) = \zeta(z)$ is an invariant of the topological conjugacy class of

---

[7]The inspiration for this zeta function is the Weil zeta function of an algebraic variety over a finite field.

f. Moreover, the zeta function contains all the information about the numbers $N_k = N_k(f)$, where N_k counts the periodic points of period k. To proceed, we need some facts from Linear Algebra.

Let A be a $n \times n$ matrix with real entries and $\bigwedge^k A$ be the induced linear map on the exterior power $\bigwedge^k \mathbb{R}^n$. In terms of the standard basis of $\mathbb{R}^n$, $\bigwedge^k A$ is represented by the matrix of all minors of A of order k. We have:

$$\overset{0}{\bigwedge} A = 1, \quad \overset{1}{\bigwedge} A = A, \quad \overset{n}{\bigwedge} A = \det(A).$$

Lemma 5.6.1. *(Liouville's Formula). Let A be a matrix. Then*

$$\det\big(\exp(A)\big) = \exp\big(\operatorname{Tr}(A)\big).$$

This is proved in [1], p. 35.

The Taylor theorem tells us,

$$\log \frac{1}{1 - z} = \sum_{k=1}^{\infty} \frac{z^k}{k}. \tag{5.2}$$

Now, setting $N_k := \det(I - A^k)$ and considering the corresponding zeta function

$$\zeta_A(z) = \exp\left(\sum_{k=1}^{\infty} \frac{N_k}{k} z^k \right),$$

we obtain the following:

Proposition 5.6.2. *For any matrix A with integer entries, we have*

$$\zeta_A(z) = \prod_{k=0}^{n} \det\left(I - z \overset{k}{\bigwedge} A \right)^{(-1)^{k+1}}.$$

Proof. We know from Lemma 3.5.17 that

$$\det(I - A) = \sum_{j=0}^{n} (-1)^j \operatorname{Tr}\left(\overset{j}{\bigwedge} A \right).$$

Therefore

$$\log \zeta_A(z) = \sum_{k=1}^{\infty} \frac{z^k}{k} N_k = \sum_{k=1}^{\infty} \frac{z^k}{k} \det(I - A^k)$$

$$= \sum_{k=1}^{\infty} \frac{z^k}{k} \sum_{j=0}^{n} \mathrm{Tr} \left(\bigwedge^j A^k \right)$$

$$= \sum_{j=0}^{n} (-1)^j \, \mathrm{Tr} \left(\sum_{k=1}^{\infty} \frac{z^k \left(\bigwedge^j A \right)}{k} \right)$$

$$= \sum_{j=0}^{n} (-1)^j \, \mathrm{Tr} \left(- \log \left(I - z \bigwedge^j A \right) \right)$$

$$= \sum_{j=0}^{n} \mathrm{Tr} \log \left(\left(I - z \bigwedge^j A \right)^{(-1)^{j+1}} \right)$$

$$= \mathrm{Tr} \sum_{j=0}^{n} \log \left(\left(I - z \bigwedge^j A \right)^{(-1)^{j+1}} \right),$$

where the last equation above comes from Lemma 5.6.1. By exponentiating we get

$$\zeta_A(z) = \prod_{j=0}^{n} \det \exp \log \left(I - z \bigwedge^j A \right)^{(-1)^{j+1}} = \prod_{j=0}^{n} \det \left(I - z \bigwedge^j A \right)^{(-1)^{j+1}}.$$

$\square$

Since all $\bigwedge^k A$ are integer matrices, the zeta function $\zeta_A(z)$ is a rational function with numerator and denominator in $\mathbb{Z}[z]$. In addition, since 0 is never a root of the denominator, the series for $\zeta_A(z)$ converges uniformly on a sufficiently small neighborhood of 0.

If A has n eigenvalues $\lambda_1, ..., \lambda_n$, then we have $\det\left(I - z \bigwedge^k A\right) = P_k(z)$, where these polynomials are $P_0(z) = 1 - z$ and

$$P_k(z) = \prod_{1 \leq l_1 < l_2 < \cdots < l_k \leq n} (1 - z\lambda_{l_1} \cdots \lambda_{l_k}),$$

with $1 \leq k \leq n$.

As we know, given a compact smooth manifold M and a smooth map $f : M \to M$, its Lefschetz number $L(f)$ is defined as

$$L(f) = \sum_{k=0}^{n} (-1)^k \operatorname{Tr}(f^{\star,k}),$$

where $f^{\star,k} : H^k(M, \mathbb{Q}) \to H^k(M, \mathbb{Q})$ are the linear maps on the cohomology spaces, $H^k(M, \mathbb{Q})$, induced by f.

Our goal is to obtain information on the set of periods of $f : M \to M$. To this end, it will be useful to have information on the whole sequence $\{L(f^m)\}_{m=0}^{\infty}$ of Lefschetz numbers of all the iterates of f. This can be done by means of the Lefschetz zeta function, $\zeta_f(z)$, as this function generates the whole sequence of Lefschetz numbers and can be independently computed as follows:

Theorem 5.6.3.

$$\zeta_f(z) = \prod_{k=0}^{n} \det\left(I_{n_k} - z f^{\star,k}\right)^{(-1)^{k+1}},$$

where $n = \dim M$, $n_k = \dim H^k(M, \mathbb{Q})$, I_{n_k} is the $n_k \times n_k$ identity matrix, and we take $\det(I_{n_k} - z f^{\star,k}) = 1$ if $n_k = 0$. In particular, $\zeta_f(z)$ is a rational function of z.

Proof. From the Lefschetz formula we know that

$$\Lambda(f^n) = \operatorname{Fix}(f^n) = \sum_{x = f^n(x)} \operatorname{Ind}_f^n(x) = L(f^n).$$

Therefore, the Lefschetz zeta function is given by

$$
\zeta_f(z) = \exp\left(\sum_{n=1}^{\infty} \frac{z^n}{n} \sum_{k=0}^{\dim M} (-1)^k \operatorname{Tr} f^{\star,k} \right)
$$

$$
= \prod_{k=0}^{\dim M} \exp\left((-1)^k \operatorname{Tr} \sum_{n=1}^{\infty} \frac{(zf^{\star,k})^n}{n} \right)
$$

$$
= \prod_{k=0}^{\dim M} \det\left(\exp(-\log(I - zf^{\star,k})) \right)^{(-1)^{k+1}}
$$

$$
= \prod_{k=0}^{\dim M} \det\left(I - zf^{\star,k} \right)^{(-1)^{k+1}}.
$$

This can be written in a more convenient form as,

$$
\zeta_f(z) = \frac{\prod_{k \, \mathrm{odd}=1}^{n} \det\left(I_{n_k} - zf^{\star,k} \right)}{\prod_{k \, \mathrm{even}=1}^{n} \det\left(I_{n_k} - zf^{\star,k} \right)}.
$$

Since all $f^{\star,k}$ are integer matrices, the numerator and the denominator are polynomials in z. Thus $\zeta_f(z)$ is a rational function with integer coefficients. □

This last fact means that all the information on the infinite sequence of integers $\{L(f^m)\}_{m=0}^{\infty}$ is contained in the two polynomials in $\mathbb{Z}[z]$ which comprise the numerator and denominator.

Remark 5.6.4. We have seen in Chapter 3 that if $L(f) \neq 0$, f has a fixed point. From the definition of the Lefschetz zeta function it follows that f is Lefschetz periodic point free if and only if

$$
\zeta_f(z) \equiv 1.
$$

In what follows we will use this essential fact as a key point in checking various situations where a diffeomorphism f has, or does not have, fixed points. These examples can be found in [49], [50] and [10].

Corollary 5.6.5. *Let $f = f_A$ be a hyperbolic toral automorphism induced by the matrix A on $\mathbb{T}^2$. Then, the induced maps on the cohomology level are*

$$f^{\star,0} = (1), \quad f^{\star,1} = A, \quad and \quad f^{\star,2} = \operatorname{sgn}(\det(A)) \cdot 1.$$

Therefore, applying the Theorem 5.6.3, if $\det(A) > 0$ we get

$$\zeta_f(z) = \frac{I - zA}{(1-z)^2}$$

and in the case where $\det(A) < 0$, we have

$$\zeta_f(z) = \frac{\det(I - zA)}{1 - z^2}.$$

We first treat a case where we know the answer.

Corollary 5.6.6. *Consider the compact manifold $M = S^1$ (the unit circle), and $f : S^1 \to S^1$, the reflection through the x-axis, i.e. $x = e^{i\theta} \mapsto f(x) = e^{-i\theta}$. Then f has Lefschetz number 2, and the second iterate f^2 is the identity map, which has Lefschetz number 0. Thus all odd iterates have Lefschetz number 2 and all even iterates have Lefschetz number 0. Therefore the zeta function of f is*

$$\zeta_f(z) = \exp \sum_{n=1}^{\infty} \frac{2z^{2n+1}}{2n+1}$$

$$= \exp 2 \left(\sum_{n=1}^{\infty} \frac{z^n}{n} - \sum_{n=1}^{\infty} \frac{z^{2n}}{2n} \right)$$

$$= \exp \left(-2 \log(1-z) + \log(1-z^2) \right) = \frac{1 - z^2}{(1-z)^2}$$

$$= \frac{1+z}{1-z}.$$

Hence, f must have fixed points.

Corollary 5.6.7. *Let* $M = S^n$, $n \geq 1$ *and let* $f : S^n \to S^n$ *be a smooth map. The cohomology groups*[8] *For* S^n *over* $\mathbb{Q}$ *these are*

$$H^k(S^n, \mathbb{Q}) = \begin{cases} \mathbb{Q} & \text{if } k \in \{0, n\}, \\ 0 & \text{otherwise.} \end{cases}$$

Now, the induced linear maps are $f^{\star,0} = (1)$, $f^{\star,i} = (0)$ *for* $i = 1, ..., n-1$ *and* $f^{\star,n} = (d)$, *where* d *is the degree of the map* f. *From (2) the Lefschetz zeta function of* f *is*

$$\zeta_f(z) = \frac{(1 - dz)^{(-1)^{n+1}}}{1 - z}.$$

So, $\zeta_f(z) \equiv 1$ *if* $d = 1$ *and* n *is odd.*

Corollary 5.6.8. *Consider a continuous map* $f : S^n \times S^n \to S^n \times S^n$, $n \geq 1$. *The cohomology groups over* $\mathbb{Q}$ *are*

$$H^k(S^n \times S^n, \mathbb{Q}) = \begin{cases} \mathbb{Q} & \text{if } k \in \{0, 2n\}, \\ \mathbb{Q} \oplus \mathbb{Q} & \text{if } k = n, \\ 0 & \text{otherwise.} \end{cases}$$

The induced linear maps are $f^{\star,0} = (1)$, $f^{\star,2n} = (g)$ *where* g *is the degree of the map* f, $f^{\star,k} = (0)$ *for any* $k \neq 1, n, 2n$, *and finally*

$$f^{\star,n} = \begin{pmatrix} a & b \\ c & d \end{pmatrix}, \quad a, b, c, d \in \mathbb{Z}.$$

Hence the Lefschetz zeta function is

$$\zeta_f(z) = \frac{\left(1 - (a+d)z + (ad - bc)z^2\right)^{(-1)^{n+1}}}{(1-z)(1-gz)}.$$

Therefore, if n *is even,* $\zeta_f(z) \neq 1$, *and if* n *is odd* $\zeta_f(z) = 1$ *if and only if* $a + d = 1 + g$ *and* $ad - bc = g$.

---

[8]Since we use cohomology with rational coefficients, cohomology groups can be replaced by the corresponding homology groups.

Thus a continuous map $f : S^n \times S^n \to S^n \times S^n$, $n \geq 1$, of degree g has no periodic points if and only if n is odd, and $a + d = 1 + g$ and $ad - bc = g$.

We now turn to complex projective space, $\mathbb{C}P^n$.

Corollary 5.6.9. *Let f be a continuous map $f : \mathbb{C}P^n \to \mathbb{C}P^n$, $n \geq 1$, with $f^{\star,2} = (a)$. Then, f is periodic point free if and only if $a = 0$.*

Proof. The cohomology groups of $\mathbb{C}P^n$ over $\mathbb{Q}$ are:

$$H^k(\mathbb{C}P^n, \mathbb{Q}) = \begin{cases} \mathbb{Q} & \text{if } k \in \{0, 2, 4, ..., 2n\}, \\ 0 & \text{otherwise.} \end{cases}$$

The induced maps are $f^{\star,k} = (a^{k/2})$, for any $k \in \{0, 2, ..., 2n\}$, with $a \in \mathbb{Z}$, and $f^{\star,k} = (0)$ otherwise. Therefore, the Lefschetz zeta function is

$$\zeta_f(z) = \left(\prod_{k=0,2,...,2n} (1 - a^{k/2}z) \right)^{-1}.$$

Hence, $\zeta_f(z) = 1$ if and only if $a = 0$. $\square$

Actually, something similar holds for the quaternionic projective space, $\mathbb{H}P^n$. For this one has to assume $f^{\star,4} = (a)$. The cohomology groups are then,

$$H^k(\mathbb{H}P^n, \mathbb{Q}) = \begin{cases} \mathbb{Q} & \text{if } k \in \{0, 4, 8, ..., 4n\}, \\ 0 & \text{otherwise} \end{cases}$$

and the induced maps are $f^{\star,k} = (a^{k/4})$ if $k \in \{0, 4, ..., 4n\}$, $a \in \mathbb{Z}$, and $f^{\star,0} = (0)$ otherwise.

Chapter 6

Borel's Fixed Point Theorem in Algebraic Groups

In order to understand the fixed point theorem of A. Borel we first discuss some properties of complete and projective varieties. Here K will denote an algebraically closed field of characteristic zero. In practice, $K = \mathbb{C}$. Topology will usually mean the Zariski topology.

6.1 Complete Varieties and Borel's Theorem

Definition 6.1.1. The variety X is complete if for all varieties Y the projection $\pi_Y : X \times Y \to Y$ is a closed map.

Lemma 6.1.2. *Some preliminary results concerning complete varieties.*

1. *A closed subvariety of a complete variety is complete.*

2. *If X is a complete variety and ϕ is a morphism, then $\phi(X)$ is closed and complete.*

3. *A finite product of complete varieties is complete.*

Proof. Let X be a complete variety, V be a closed subvariety and Y an arbitrary variety.

1. Since $\pi_Y : X \times Y \to Y$ is closed, if C is a Zariski closed subset of V, then C is also closed in X. Hence $\pi_Y(C)$ is closed. Thus for every Y, the projection $V \times Y \to Y$ is a closed map.

2. Let X and Y be complete varieties and $\phi : X \to \phi(X) \subseteq Y$ be a morphism. Because ϕ is Zariski continuous its graph, $\{(x,y) : x \in X,\ y = \phi(x)\}$, is *closed* in $X \times Y$. Since X is complete, $\phi(X)$ is Zariski closed in Y. Hence by 1) $\phi(X)$ is complete because Y is.

3. Let $V = \prod_{i=1}^{n} X_i$ be a finite product of complete varieties. We show $\prod_{i=1}^{n} X_i$ is complete by induction on n. Let Y be an arbitrary variety. By induction we know $\prod_{i \geq 2}^{n} X_i \times (X_1 \times Y) \to (X_1 \times Y)$ is a closed map. That is, $\prod_{i \geq 1}^{n} X_i \times Y \to X_1 \times Y$ is a closed map. But since X_1 is itself complete, the projection of $X_1 \times Y$ onto Y is also a closed map. When we compose these two we indeed get a closed map $\prod_{i=1}^{n} X_i \times Y \to Y$.

$\square$

Lemma 6.1.3. *The field K is not complete.*

Proof. Suppose K were complete. Let $Y = K$ and $A = \{(x,y) \in K \times K : xy = 1\}$. Then A is a closed subvariety of $K \times K$ and so is also complete by Lemma 6.1.2. But $\pi_Y(K) = K^\times$ which is closed. It is also open. Since K is connected this is a contradiction. $\square$

The purpose of all this is the following:

Corollary 6.1.4. *A complete connected affine variety X is just a point.*

Proof. Since X is affine $X \subseteq K^n$. For $i = 1, \ldots, n$, let ϕ_i be the projection onto the ith coordinate. Since these maps are affine morphisms each $\phi_i(X)$ is a connected complete affine subvariety of K by Lemma

6.1.2. If a ϕ_i maps X onto K, Then K would be a complete variety, which contradicts Lemma 6.1.3. Hence none of the ϕ_i is onto so each $\phi_i(X)$ is a proper affine subvariety of K which has dimension 1. Thus each $\phi_i(X)$ has dimension 0 and so is finite. Since X is connected so is $\phi_i(X)$. Hence $\phi_i(X)$ must be a point for each i and so X itself is a point. $\square$

We can now state the Borel fixed point theorem. As the reader can see, in applying this result the issue will be knowing that X is a complete variety.

Theorem 6.1.5. *Let G be a Zariski connected solvable algebraic group over K and X be a nonempty complete variety over K. Then X has a G-fixed point.*

Proof. The proof is by induction on the dimension d of G. If $d = 0$, then $G = (1)$ and so the result is trivially true. Assume $d > 0$. Then $N = [G, G]$ is a normal connected subgroup of G which, since it is solvable, must be of lower dimension. Hence by induction, N has a fixed point. Let $\mathcal{F}$ be the set of these fixed points. $\mathcal{F}$ is clearly closed in X and hence is also a complete variety. If $x \in \mathcal{F}$ and $g \in G$ then $gN(x) = g(x)$ since $\mathcal{F}$ is N-fixed. But because N is normal in G this means $Ng(x) = g(x)$. Thus $g(x)$ is also N fixed so $g(x) \in \mathcal{F}$. This means G leaves $\mathcal{F}$ stable. By the closed orbit lemma there exists an $x_0 \in \mathcal{F}$ so that $\mathcal{O}_G(x_0)$ is closed. Let G_{x_0} be the isotropy group of x_0. Then $N \subseteq G_{x_0}$ and so G_{x_0} is also normal in G. Thus the natural map $G/G_{x_0} \to \mathcal{O}_G(x_0)$ is a bijective morphism of varieties. The left side is affine since G and G_{x_0} are normal. Because it is closed the right side is complete. Hence so is the left side. By Lemma 6.1.4 G/G_{x_0} and therefore also $\mathcal{O}_G(x_0)$ is a point and so x_0 is G-fixed. $\square$

Although not needed, for its interest we include the following Proposition as it is the *holomorphic* analogue of Borel's fixed point theorem. It depends on *Liouville's theorem*. Here we use the Euclidean topology.

Proposition 6.1.6. *The action of a complex connected solvable Lie group G operating holomorphically on a compact complex manifold X is trivial.*

Proof. Exactly as in the proof of Borel's theorem, by induction, $[G, G]$ fixes X. Since X is compact $\mathcal{O}_G(x)$ has compact closure for each $x \in X$. Hence the orbit map ϕ taking $g \mapsto g \cdot x$ is holomorphic and bounded. As such $\phi \cdot \exp$ taking $\mathfrak{g} \to G_x$ is also holomorphic and bounded and so constant by Liouville's theorem. But since G is a complex group, $\exp(\mathfrak{g})$ is dense in G so that $\phi(G) = G_x$ is a point which must be x. ☐

In order to use Theorem 6.1.5 to prove the global version of Lie's theorem (the Lie-Kolchin theorem) we shall need to know the flag variety,

$$\mathcal{F}(V) = \mathrm{GL}(n, K)/T(n, K),$$

is complete, where $T(n, K)$ denotes the full group of upper triangular matrices. We will prove this in the next section. Then we will use Lie's theorem to prove that actually, G/B is complete for any connected (affine) linear algebraic group G and Borel subgroup B (Corollary 6.4.4).

6.2   The Projective and Grassmann Spaces

Let V be a finite dimensional vector K-space and $P(V)$ the associated projective space. That is, the set of all lines through 0 in V. If V has dimension n, then $P(V)$ has dimension $n - 1$. Another way to describe $P(V)$ is the set of equivalence classes $[v]$ of nonzero vectors $v \in V$, where $[v] = [w]$ means that $v = tw$ for some nonzero $t \in K^{\times}$. Let $\pi : V \setminus (0) \to P(V)$ be the projection given by $\pi(v) = [v]$. and topologize $P(V)$ so that π is continuous and open, where $V \setminus (0)$ is viewed as an open set in V and V takes the Zariski topology. Thus $U \subseteq P(V)$ is open if and only if $\pi^{-1}U \subseteq V$ is open.

For $g \in GL(V)$ define $g^- : P(V) \to P(V)$ by $g^-(v^-) = g(v)^-$. As explained above, g^- is well defined, $g^- \pi = \pi g^-$, $(gh)^- = g^- h^-$ and if $\lambda \neq 0$, $(\lambda g)^{-} = g^{-}$. Now, suppose one has a linear representation $G \times V \to V$ of G on V. Then this induces a compatible action of G on $P(V)$, making the diagram below commutative.

$$
\begin{array}{ccc}
G \times (V \setminus (0)) & \longrightarrow & V \setminus \{0\} \\
\downarrow \pi & & \downarrow \pi \\
G \times P(V) & \longrightarrow & P(V)
\end{array}
$$

Let $\mathcal{G}^r(V)$, the Grassmann space, be the set of all subspaces of V of fixed dimension r. For example, $\mathcal{G}^1(V)$ and $\mathcal{G}^{n-1}(V)$ are, or are isomorphic to $P(V)$. Now just as above $GL(V)$ also acts on $\mathcal{G}^r(V)$. This action is transitive. For if U and W are subspaces of V of dimension r, take a basis $w_1, \ldots, w_r$ and $u_1, \ldots, u_r$ of each and then extend these each to bases $w_1, \ldots, w_n$ and $u_1, \ldots, u_n$ of V. The linear transformation T taking each w_i to u_i is invertible and $T(W) = U$. Since $GL(V)$ acts transitively on $\mathcal{G}^r(V)$ with $\mathrm{Stab}_{GL(V)}(W)$, $GL(V)/\mathrm{Stab}_{GL(V)}(W) = \mathcal{G}^r(V)$. Now (see [1], pp. 22-23) the stabilizer of W is the set of operators of the form,

$$
\begin{pmatrix} A & B \\ 0 & C \end{pmatrix},
$$

where A is an $r \times r$ matrix and C is $(n - r) \times (n - r)$ and both are invertible as we are in $GL(V)$.

It is often useful to be able to pass from $\mathcal{G}^r(V)$ to a projective space. This can be done by the *canonical imbedding*,

$$
\phi : \mathcal{G}^r(V) \to P(\wedge^r V),
$$

which is defined as follows: For an r-dimensional subspace W of V, choose a basis $\{w_1, \ldots, w_r\}$. Then $w_1 \wedge \cdots \wedge w_r$ is a nonzero element of $\wedge^r V$ and so the line through it gives a point in $P(\wedge^r V)$.

The following Lemma is a well known fact of multilinear algebra which we leave to the reader.

Lemma 6.2.1. *Let* $\{w_1, \ldots, w_r\}$ *and* $\{u_1, \ldots, u_r\}$ *each be* r *linearly independent vectors in* V. *Then* $w_1 \wedge \cdots \wedge w_r = \lambda u_1 \wedge \cdots \wedge u_r$ *for* $\lambda \neq 0 \in K$ *if and only if they both have the same* K-*linear span,* W. *(The* λ *being* $\det C$, *where* C *is the* K-*linear map of* W *taking* w_i *to* u_i, $i = 1, \ldots, r$.*)*

This leads us immediately to:

Proposition 6.2.2. *The map* $\phi : \mathcal{G}^r(V) \to P(\wedge^r V)$ *is well defined and injective.*

Remark 6.2.3. This map is not in general surjective. For if it were, it would be a morphism of connected varieties which would therefore have the same dimension. But

$$\dim P(\wedge^r V) = \frac{n!}{r!(n-r)!},$$

and since $\mathrm{GL}(V)/\mathrm{Stab}_{\mathrm{GL}(V)}(W) = \mathcal{G}^r(V)$ and we know $\mathrm{Stab}_{\mathrm{GL}(V)}(W)$, we can calculate its dimension. This is, $\dim \mathrm{Stab}_{\mathrm{GL}(V)}(W) = n^2 - (n - r)n$, while $\dim \mathrm{GL}(V) = n^2$ and so $\mathcal{G}^r(V)$ has dimension $r(n-r)$. It follows $\mathcal{G}^r(V)$ is morphicly equivalent with $P(\wedge^r V)$, only if $r = 1$ or $r = n - 1$. That is, only when $\mathcal{G}^r(V)$ is isomorphic with $P(V)$.

Proposition 6.2.4. *Let* ρ *be a morphic representation of* G *on* V. *Then this gives rise to a morphic equivariant map for the induced action of* G *on* $\mathcal{G}^r(V)$.

Proof. Since $\mathrm{GL}(V)$ acts transitively and continuously on $\mathcal{G}^r(V)$, the latter is a quotient space $\mathrm{GL}(V)/\mathrm{Stab}_{\mathrm{GL}(V)}(W)$, where W is some fixed r-dimensional subspace of V. If $\gamma : \mathrm{GL}(V) \to \mathcal{G}^r(V)$ denotes the corresponding projection and $\{w_1, \ldots, w_r\}$ is a basis of W, then since $\{gw_1, \ldots, gw_r\}$ are linearly independent for each $g \in G$, $g \mapsto gw_1 \wedge \cdots \wedge gw_r$ is a map $\psi : \mathrm{GL}(V) \to \wedge^r V - (0)$. Clearly, ϕ factors as $\phi_1 \pi$, where $\phi_1 : \mathcal{G}^r(V) \to \wedge^r V - (0)$ and $\pi : \wedge^r V - (0) \to P(\wedge^r V)$ is the natural map. The diagram below, including the map ϕ_1 is commutative, since $\phi_1 \gamma(g) = \phi_1(gW) = gw_1 \wedge \cdots \wedge gw_r = \psi(g)$.

$$\begin{CD} \mathrm{GL}(V) @>{\psi}>> \wedge^r V \setminus \{0\} \\ @V{\gamma}VV @VV{\pi}V \\ \mathcal{G}^r(V) @>>{\phi}> P(\wedge^r V) \end{CD}$$

To see that ϕ is a morphism, it is enough to show ϕ_1 is and hence that ψ is because γ and π are both morphisms. But, clearly, ψ is a morphism since $g \mapsto gw_1 \otimes \cdots \otimes gw_r$ is. Moreover ϕ intertwines the actions.

$$\begin{CD} \mathcal{G}^r(V) @>{\phi}>> P(\wedge^r V) \\ @V{g}VV @VV{(g\wedge\ldots\wedge g)^-}V \\ \mathcal{G}^r(V) @>>{\phi}> P(\wedge^r V) \end{CD}$$

For, let $h(W)$ be any point of $\mathcal{G}^r(V)$ and $g \in G$. Then $\phi(gh(W)) = (ghw_1 \wedge \cdots \wedge ghw_r)^-$. While

$$\begin{aligned}
(g \wedge \cdots \wedge g)^-(\phi(h(W))) &= (g \wedge \cdots \wedge g)^-(hw_1 \wedge \cdots \wedge hw_r) \\
&= [(g \wedge \cdots \wedge g)(hw_1 \wedge \cdots \wedge hw_r)]^- \\
&= (ghw_1 \wedge \cdots \wedge ghw_r)^-.
\end{aligned}$$

$\square$

6.3 Projective Varieties

Definition 6.3.1. Any closed subvariety, X, of $P(V)$ is called a *projective variety*.

Proposition 6.3.2. *We have the following:*

1. *A product of projective varieties is a projective variety.*

2. *Projective varieties are complete.*

3. *In particular, the Grassmann variety and flag variety are projective and so are complete.*

Proof. To prove the product of P_n and P_m is a projective variety we define the *Segre imbedding*,

$$s : P_n \times P_m \to P_{nm+n+m},$$

as follows: Let $x = (x_0, \ldots, x_n)$ be homogeneous coordinates in P_n and $y = (y_0, \ldots, y_m)$ be homogeneous coordinates in P_m. Then $s(x, y)$ will be a point with $(n + 1)(m + 1)$ homogeneous coordinates given by $s(x, y) = x_i y_j$, where $i = 0, \ldots, n$ and $j = 0, \ldots, m$. Hence $s(x, y)$ is a point in P_{nm+n+m}. If U_i is the affine open set of K^n defined by $x_i \neq 0$ and V_j is the affine open set of K^m defined by $y_j \neq 0$, then these affine open sets cover K^n and K^m respectively. Let $W_{i,j}$ be the affine open set in K^{nm+n+m}, where $z_{i,j} \neq 0$. Then $s(U_i \times V_j) = W_{i,j}$ and is clearly a morphism onto $W_{i,j}$. Also, $s^{-1}(W_{i,j}) = U_i \times V_j$ so the restriction of s to each $U_i \times V_j$ is injective and so s itself is injective and evidently the inverse map $W_{i,j} \to U_i \times V_j$ is also a morphism. Thus $s(P_n \times P_m)$ is a closed subvariety of P_{nm+n+m}.

To prove 1 and 2 it suffices by Lemma 6.1.2 to show projective space itself is complete. However, doing so is not simple and requires some algebraic geometry and commutative algebra and for this we refer to [97], Theorem 13.4.5, p. 177.

For 3 (which is what we need) we now show the Grassmann and flag varieties are projective varieties. We turn first to the Grassmann variety. As we know the map $\phi : \mathcal{G}^r(V) \to P(\wedge^r V)$ is injective. We now show the image of $\mathcal{G}^r(V)$ is Zariski closed making $\mathcal{G}^r(V)$ a projective variety. Let $v_1, \ldots, v_n$ be a fixed basis of V and cover $P(\wedge^r V)$ by a finite number of open sets U as follows. Each U consists of those points whose homogeneous coordinates in the associated basis of $\wedge^r V$ have a nonzero coefficient of $v_1 \wedge \cdots \wedge v_r$. Thus U is the complement of a linear subspace of $\wedge^r V$. Write $V = W_0 \oplus W_0'$, where W_0 is spanned by $v_1, \ldots, v_r$ and W_0' by $v_{r+1}, \ldots, v_n$. Then for $W \in \mathcal{G}^r(V)$, $\phi(W) \in U$ if and only if the projection onto W_0 maps W onto isomorphically onto W_0. In this case

W has a unique basis $v_1 + x_1(W), \ldots, v_r + x_r(W)$, where $x_i(W) \in W_0'$ for $i = 1, \ldots, r$. Then $x_i(W) = \sum_{j>r} a_{i,j} v_j$. This means $\phi(W)$ is the projection into $P(\wedge^r V)$ of the vector

$$v + \Big(\sum_{1 \le i \le r} v_1 \wedge \cdots \wedge x_i(W) \wedge \cdots \wedge v_r \Big) + *,$$

where $*$ involves basis vectors omitting 2 or more of the $v_1, \ldots, v_r$. The term in parenthesis equals

$$\sum_{j>r} a_{i,j} v_1 \wedge v_j \wedge v_r.$$

In this way for $1 \le i \le r$ and $j > r$, in $\big(v_1 + x_1(W)\big) \wedge \cdots \wedge \big(v_r + x_r(W)\big)$, we recover $a_{i,j}$ as the coefficient of the basis vector $v_1 \wedge v_j \wedge v_r$ and these coefficients, which determine W, are arbitrary while the coefficients of the remaining vectors in $\wedge^r(V)$ are polynomial functions of these so that $\phi(\mathcal{G}^r(V))$ is essentially the graph of a morphism from the linear space of the $a_{i,j}'s$ to some other linear space and so it is Zariski closed. Thus each $\mathcal{G}^r(V)$ is closed in $P(\wedge^r V)$ and so a projective variety.

Turning to the flag variety, we observe that if $W \in \mathcal{G}^r$ and $W' \in \mathcal{G}^{r'}$, where $r \le r'$, the condition $W \subseteq W'$ can be expressed by algebraic equations in the coordinates of $P(\wedge^r V) \times P(\wedge^{r'} V)$. Hence

$$\{(W, W') \in \mathcal{G}^r \times \mathcal{G}^{r'} : W \subseteq W'\}$$

is a closed subvariety. Intersecting a finite number of these closed sets tells us for $n = \dim V$ that

$$\mathcal{F}(V) = \{(V_1, \ldots, V_n) \in \mathcal{G}^1(V) \times \mathcal{G}^2(V) \times \ldots \times \mathcal{G}^n(V) : V_i \subseteq V_{i+1}, \; i < n\}$$

is also a Zariski closed set and so $\mathcal{F}(V)$ is a projective variety as well. $\square$

Since to apply the Borel fixed point theorem what we really need to know is just that $\mathcal{G}^r(V)$ and $\mathcal{F}(V)$ are complete, here is an alternative approach dealing directly with these varieties.

Lemma 6.3.3. $\mathrm{GL}(n, K)$ *is Zariski connected.*

Proof. First we observe that $\mathrm{GL}(1, K) = K^\times$ is Zariski connected. For suppose $K^\times$ were the disjoint union of two Zariski closed sets. Since here we are dealing with polynomials in 1 variable and these have a finite number of roots this would mean $K^\times$ is finite, a contradiction since K has characteristic zero. Let D_n be the diagonal elements of $\mathrm{GL}(n, K)$. Then D_n is a product of n copies of $\mathrm{GL}(1, K)$ and as such is also Zariski connected. Hence so is $g D_n g^{-1}$ for every $g \in \mathrm{GL}(n, K)$. But since each of these conjugates contains the identity their union D is also Zariski connected as is its Zariski closure. Now D is the set of all diagonalizable elements in $\mathrm{GL}(n, K)$. As in the proof of Lemma 3.5.17, taking into account that the eigenvalue zero is absent, D is Zariski dense in $\mathrm{GL}(n, K)$. Hence the conclusion.   □

The following works for any connected algebraic subgroup $G \subseteq \mathrm{GL}(V)$. However we will work with $G = \mathrm{GL}(V)$ itself. Define a *course flag*,

$$V_0 \subset V_1 \subset \ldots \subset V_s,$$

where here we do not require V_i/V_{i-1} to have dimension 1. Then the (algebraic) subgroup P of G which preserves this course flag is called a *parabolic subgroup* of G. Evidently, any parabolic subgroup of G contains a Borel subgroup. Indeed, since a course flag can be refined to a flag, each Borel subgroup is a minimal parabolic subgroup. An important property of parabolic subgroups P of connected algebraic groups G is that the variety G/P is complete. This is because G/B is complete (see Corollary 6.4.4 below) and since $B \subseteq P$, the variety G/B maps onto G/P so it is also complete. For further details on Borel subgroups see the next section.

Since, as we know, $\mathrm{GL}(n, K)$ acts transitively on $\mathcal{G}^r(V)$ with the stabilizer of W the operators of the form given above, the stabilizer is a parabolic subgroup and so $\mathcal{G}^r(V)$ is complete. Of course, similarly the flag variety, $\mathcal{F}(V) = \mathrm{GL}(n, K)/T(n, K)$, is also complete.

We conclude this section with the following:

Remark 6.3.4. Now $X = P(V)$, and indeed $\mathcal{G}^r(V)$ and $\mathcal{F}(V)$ are compact in the Euclidean topology (see [1], pp. 22-23, which applies to the real case as well). However, this will not help in proving these are projective varieties and therefore complete. For here is how such a "proof" would have to go: Let Y be an arbitrary variety and C be a Zariski closed set in $X \times Y$. Then C is also a closed set in the Euclidean topology since polynomials are Euclidean continuous and therefore the intersection of zero sets of polynomials are Euclidean closed. By compactness of X, $\pi_Y(C)$ is Euclidean closed in Y. But what we need is that it is *Zariski* closed in Y and it is precisely this implication, and with it the equivalence of the two topologies, that is false. For example, the Euclidean topology is Hausdorf, but the Zariski topology is not.

6.4   Consequences of Borel's Fixed Point Theorem

As a corollary we get Lie's theorem.

Corollary 6.4.1. *Let ρ be a morphic representation of the connected complex solvable group G on the complex vector space V. Then there is a G-stable flag on V. That is, $\rho(G)$ can be put in triangular form.*

Proof. Evidently we can replace G by $\rho(G)$ and call that the new G. By Lemma 6.3.2, or the remarks just above, the flag variety on V is complete. Since G is a connected solvable complex algebraic group, Borel's fixed point theorem tells us there is a G-fixed point in the flag variety under the induced action. But then, by the remarks in the Introduction, this is a G-stable flag on V.   □

Corollary 6.4.2. *Let G be any solvable subgroup of $\mathrm{GL}(V)$. Then some finite index subgroup of G can be put in triangular form.*

Proof. Let $G^{\#}$ be the hull of G. Then $G^{\#}$ is a solvable algebraic subgroup of $\mathrm{GL}(V)$ as is its Zariski connected component $(G^{\#})_0$. By Lie's theorem $(G^{\#})_0$ can be put in triangular form. Now $(G^{\#})_0$ has finite index in $G^{\#}$. Therefore $G \cdot (G^{\#})_0/(G^{\#})_0$ is also finite. But $G \cdot (G^{\#})_0/(G^{\#})_0 \cong G/G \cap (G^{\#})_0$. Hence $G \cap (G^{\#})_0$ has finite index in G and is in triangular form.   □

Definition 6.4.3. A Borel subgroup B of a linear algebraic group G is a maximal connected solvable subgroup of G.

Since each $\{\exp(zX) : z \in K\}$ is connected and abelian and therefore solvable, for dimension reasons Borel subgroups exist. Of course if B is a Borel subgroup and α is an automorphism of G, then $\alpha(B)$ is also a Borel subgroup. For example, when $G = \mathrm{GL}(n, K)$, a Borel subgroup is the full (upper or lower) triangular group, $T(n, K)$. (The map $g \mapsto (g^t)^{-1}$ is an automorphism taking the upper to the lower triangular matrices in $\mathrm{GL}(n, K)$). This is because $T(n, K)$ is clearly connected and solvable. It is also a maximal such. For suppose B were a strictly larger such group. By Lie's theorem 6.4.1 there is a basis so that $B \subseteq T(n, K)$. Hence $\dim B \leq \dim T(n, K)$, a contradiction since $B \supset T(n, K)$.

A final consequence of the Borel fixed point theorem is the following which gives yet another proof G/B is a projective variety.

Corollary 6.4.4. *Let G be any connected linear algebraic group. Then its Borel subgroups B are all conjugate and for any B, G/B is a projective variety.*

Proof. Consider a Borel subgroup B_* of G of maximal dimension. We first show G/B is a projective variety. A basic theorem of algebraic groups tells us there is a faithful morphic representation $\rho : G \to \mathrm{GL}(V)$ and a line V_1 in V so that $B_* = \mathrm{Stab}_G(V_1)$. The Lie Kolchin theorem applied to the representation gotten from ρ of G on V/V_1 yields a B_* stable flag $F = (V_1, \ldots, V_n)$ on V. Now G operates on $\mathcal{F}(V)$ and this action gives a map $G/B_* \to G(F)$. This map is an isomorphism of varieties because the map $G/B_* \to G(V_1) \subseteq P(V)$ is already an isomorphism of varieties.

Now suppose B is any Borel subgroup of G. By Lie's theorem, B stabilizes a flag. Since $\dim B \leq \dim B_*$, $\dim G/B \geq \dim G/B_*$. That is, the G-orbit of F has minimal dimension among of all G-orbits of flags in V. The closed orbit lemma tells us the G-orbit of F is closed. Hence G/B_* is a projective variety.

Now let B again be any other Borel subgroup. Let B operate on G/B_* by left translation. Since B is solvable and connected and G/B_*

is a projective variety and therefore complete, by the Theorem 6.1.5, there is a B-fixed point gB_*. Since $B(gB_*) = gB_*$, there is a $g \in G$ so that $g^{-1}Bg \subseteq B_*$. But since $g^{-1}Bg$ is certainly a Borel subgroup and B_* is maximal, $g^{-1}Bg = B_*$. This shows they are all conjugate and so also all G/B are projective varieties. $\square$

We conclude this section with an application of fixed point theory involving the Grassmann space and subgroups H of G of cofinite volume.

In the course of generalizing the well known density theorem of Borel it was necessary to consider a vector space V over k of finite dimension, n, where $k = \mathbb{R}$, or more generally, any subfield of $\mathbb{C}$. $GL(V)$ denotes, as usual, the general linear group of V and $P(V) = \mathcal{G}^1(V)$ is projective space. Each $\mathcal{G}^r(V)$ is a compact manifold and $\mathcal{G}(V)$, the Grassmann space of V, is a disjoint union of these open submanifolds. As above, $\pi : V \setminus (0) \to P(V)$ denote the canonical map $v \mapsto [v]$. If W is a subspace of V of dimension ≥ 1, then $[W]$ denotes the corresponding subvariety of $P(V)$. An action of G on V gives rise to one on $\mathcal{G}(V)$ and a G-invariant subspace in V is a G-fixed point in this new action. Then one has,

Theorem 6.4.5. *Let G be a Lie group satisfying certain appropriate hypotheses (see [1], p. 386), ρ a smooth representation of G on V and H is a closed subgroup of G with G/H of finite volume. Then each H-invariant subspace of V is G-invariant.*

Proof. Suppose the dimension of the H-invariant subspace W is r. Form $\mathcal{G}^r(V)$ and consider the action $G \times \mathcal{G}^r(V) \to \mathcal{G}^r(V)$. Here W corresponds to a point $p \in \mathcal{G}^r(V)$. Since H leaves W stable, p is H-fixed. So $H \subseteq \mathrm{Stab}_G(p)$. Now because G/H has a finite G-invariant measure, the same is true of $G/\mathrm{Stab}_G(p)$. This means that $p \in \mathrm{Supp}\,\mu$ for an appropriate G-invariant finite positive measure μ on $\mathcal{G}^r(V)$, where $\mathrm{Supp}\,\mu = G \circ p$. Using these "certain hypotheses" one can show that p is actually a G-fixed point and hence W is G-stable. $\square$

6.5   Two Conjugacy Theorems for Real Linear Lie Groups

In this final section we sketch the proof of two fixed point theorems whose proofs were given very recently in [8] and which also result in conjugacy theorems. These results, concerning real reductive linear groups, seem very intuitive and the authors agree with those of [8] that they deserve to be better known. They have somewhat the same flavor as the Borel fixed point theorem and are originally due, independently, to Mostow and Vinberg.

Theorem 6.5.1. *Let G be a real reductive Lie subgroup of $\mathrm{GL}(n, \mathbb{C}) = \mathrm{GL}(V)$ and $G = KAN$ be an Iwasawa decomposition of G. Then each unipotent subgroup U of G is conjugate by something in G into N. In particular, any two maximal unipotent subgroups of G are conjugate.*

Proof. Let $H = (AN)^{\#}$, the hull of AN. Then $[H : AN] < \infty$ and since G is linear therefore K is compact, G/H is also compact. We let U act on the manifold G/H by $u \circ (gH) = gug^{-1}H$. As shown in [8], G can be regarded as a real algebraic group so G/H is a variety and also the action has a closed orbit. One checks easily that this is a morphic action. This closed orbit, $g_0 U g_0^{-1} H/H$, is therefore compact. But because U is unipotent the orbit is a cell, that is, is diffeomorphic to some Euclidean space. Hence it is a point. *This is the fixed point theorem.* Taking $u = 1$ shows this point is H. Therefore, $g_0 U g_0^{-1} \subseteq H$. As a conjugate of U it is connected and so $g_0 U g_0^{-1} \in AN$. Also, as a conjugate of U it consists of unipotent elements and hence $g_0 U g_0^{-1} \subseteq N$. □

Similarly,

Theorem 6.5.2. *Let G be as in Theorem 6.5.1 and S be a connected solvable subgroup of G with all real eigenvalues. Then S is conjugate by something in G to AN. In particular, any two maximal connected solvable subgroups of G both of which have only real eigenvalues are conjugate.*

Proof. Let G and H be as before and replace S by the real points of its hull which is also a connected solvable subgroup of G with only real eigenvalues. Using the same action as above, the orbit $S \circ (gH) = S/\operatorname{Stab}_S(gH)$. Because the eigenvalues of S are all positive the real algebraic subgroups of S are all connected. In particular, $\operatorname{Stab}_S(gH)$ is connected and so the orbit is also a cell and as before it must be a point and therefore $g_0 S g_0^{-1} \subseteq AN$. $\square$

Chapter 7

Miscellaneous Fixed Point Theorems

7.1 Applications to Number Theory

7.1.1 The Little Fermat Theorem

Here we prove the little Fermat theorem[1].

Theorem 7.1.1. *(Little Fermat Theorem.) For any integer $a \geq 2$ and any prime p,*
$$a^p \equiv a \bmod(p).$$

To do this we make the following construction. For any integer $a \in \mathbb{Z}^+$ we consider the map

$$f_a : [0, 1] \to [0, 1],$$

defined by

$$f_a(x) = \begin{cases} ax & \text{if } 0 \leq x \leq \frac{1}{a}, \\ ax - j & \text{if } \frac{j}{a} < x \leq \frac{j+1}{a}, \end{cases}$$

where $1 \leq j \leq a - 1$.

---
[1]This proof can be found in [39].

To illustrate this construction consider the case $a = 4$. The graph of $f_4(x)$ is given below.

In order to count fixed points we need the following lemma. Here $N_m(f)$ denotes the number of points of minimal period m of f.

Lemma 7.1.2. *Let x be a point of period n for f_a. Then,*

1. *If x has minimal period m, then $m \mid n$.*

2. *Two minimal m-cycles are either distinct or identical.*

3. *For $m \geq 1$, $m \mid N_m(f)$ whenever $N_m(f)$ is finite.*

Proof. We leave the easy proofs of 1 and 2 to the reader. Concerning 3, observe that the points of minimal period m are partitioned into disjoint m-cycles. Since any m-cycle contains exactly m distinct points and there is a finite number of cycles, we get $m \mid N_m$.   □

Proposition 7.1.3. *The map f_a*

1. *Has exactly a^n points of period n.*

2. *For all $a > 1$ and $n \geq 1$*

$$a^n = \sum_{m \mid n} N_m(f_a).$$

Proof. Concerning the first item, the points of period n are exactly the fixed points of the function f_a^n. Now,

$$f_a^n(x) = \begin{cases} a^n x & \text{if } 0 \le x \le \frac{1}{a^n}, \\ a^n x - j & \text{if } \frac{j}{a^n} < x \le \frac{j+1}{a^n}, \end{cases}$$

where $1 \le j \le a^n - 1$. This shows that the graph of $(f_a)^n$ consists of a^n parallel line segments each of slope a^n. Therefore, the diagonal Δ intersects this graph in a^n distinct points and thus we have a^n fixed points. Therefore f_a has a^n distinct n-periodic points. By the Lemma 7.1.2, any point of period n is a point of minimal period m for some $m \mid n$. The second item is a direct application of 3 of the Lemma 7.1.2. □

Proof of Theorem 7.1.1.

By Proposition 7.1.3, since p is prime $A^p = N_1 + N_p = a + N_p$. Therefore, $a^p - a = N_p$, which by Lemma 7.1.2 is divisible by p.

7.1.2   Fermat's Two Squares Theorem

Our next result is a proof, using fixed points, of the well known theorem of Fermat which states that any number n all of whose prime divisors are of the form $4k + 1$ is the sum of *two squares*. These conditions are necessary since the prime $7 = 4 \cdot 1 + 3$ can only be written as a sum of 4 squares, while 21 which is of the form $4 \cdot 5 + 1$, can only be written as a sum of 3 squares. By induction on the number of primes in n, the simple identity $|z|^2|w|^2 = |zw|^2$ involving the complex numbers z and w, reduces this question to proving this for a prime p of the form $4k + 1$ and this is what we shall do.

We note that this venerable and famous result has more than 50 proofs. Here we will follow Heath-Brown ([53]) and Zagier ([104]).

Theorem 7.1.4. *Any prime p of the form $4k + 1$ is the sum of two squares.*

Proof. Let

$$A = \begin{pmatrix} 0 & 1 & 0 \\ 1 & 0 & 0 \\ 0 & 0 & -1 \end{pmatrix}, \quad B = \begin{pmatrix} 0 & 1 & 0 \\ 1 & 0 & 0 \\ 0 & 0 & 1 \end{pmatrix},$$

$$C = \begin{pmatrix} 1 & -1 & 1 \\ 0 & 1 & 0 \\ 0 & 2 & -1 \end{pmatrix}, \quad M = \begin{pmatrix} 0 & 2 & 0 \\ 2 & 0 & 0 \\ 0 & 0 & 1 \end{pmatrix}.$$

One checks easily that $A^2 = B^2 = C^2 = I$ so that each of these is its own inverse and that $A^t M A = B^t M B = C^t M C$. For the moment let p be any prime. Next, we define the following sets:

$$S_A = \{v := (x, y, z) \in \mathbb{Z}^{+3} \ : \ v^t M v = p, \ x, \ y > 0\},$$

$$S_B = \{v \in S_A \ : \ z > 0\}, \text{ and } S_C = \{v \in S_A \ : \ x + z > y\}.$$

From the fact that $v^t M v = p = 4xy + z^2$ and since $x, \ y > 0$, we see that S_A is a finite set. One checks easily that A sends S_A to itself, and similarly $B(S_B) = S_B$ and $C(S_C) = S_C$. Moreover, since p is prime,

$$S_A = S_B \cup A(S_B) \text{ and } S_A = S_C \cup A(S_C) \text{ (disjoint union)}.$$

Now, since A is a $1 : 1$ map we have

$$|S_B| = |A(S_B)| \text{ and } |S_C| = |A(S_C)|,$$

so that

$$|S_A| = 2|S_B| \text{ and } |S_A| = 2|S_C|.$$

In particular $|S_B| = |S_C|$. Since $C^2 = I$ the action of C on S_C produces orbits of cardinality 1 or 2. If (x, y, z) is a fixed point of C under this action, then we must have $x - y + z = x$, $y = y$, and $2y - z = z$. In particular $y = z$ and since $4xy + z^2 = p$, we get $p = y(4x + y)$.

Now, suppose that in addition p is also of the form $4k + 1$. Then, the last equality forces $y = 1$ and $x = \frac{p-1}{4}$. For otherwise, $y = p$ and $4x + y = 1$ which implies $x = -k$, where k is any positive number, a contradiction. This shows that C has exactly one fixed point in its action on S_C. Therefore $|S_C|$ is odd. Now, consider the action of B on

S_B. Since $|S_B|$ is therefore also odd, B must also have an fixed point. But a fixed point of B must have $x = y$. Therefore

$$p = (2x)^2 + z^2.$$

$\square$

7.2   Fixed Points in Group Theory

Here we prove a theorem of Jordan concerning the existence of fixed points when there is a transitive faithful group action (which itself has interesting applications to number theory and topology). These things can all be found in [91].

First, some notation. Let $G \times X \to X$ be a group action of a finite group G acting on a finite set X. We denote the order of G by $|G|$. For a numerical function f on G, we write

$$\int_G f = \frac{1}{|G|} \sum_{g \in G} f(g),$$

where the integral, $\int_G$, is clearly a linear functional. For a subset $S \subseteq G$, $\int_S f$ stands for $\frac{1}{|G|} \sum_{g \in S} f(g)$.

Our objective is,

Theorem 7.2.1. *If G has at least two points and acts transitively and faithfully on X, then there exists a $g \in G$ without fixed points.*

We begin with a lemma of Burnside.

Lemma 7.2.2. *For a fixed $g \in G$ let $f(g)$ be the number of points of X fixed by g. Then, the number of G-orbits of (G, X) is $\int_G f$.*

Proof. First consider the case when we have a transitive action. Then $X = G/\operatorname{Stab}(x_0)$. Let Ω be the subset $G \times X$ consisting of those (g, x), where $g(x) = x$. We compute $|\Omega|$ in two ways by the Fubini theorem. Projecting onto G, each fiber over $g \in G$ has $f(g)$ elements so $|\Omega| = \sum_{g \in G} f(g)$. On the other hand projecting onto X, each fiber over $x \in X$

is a conjugate of $\mathrm{Stab}_G(x_0)$ and so has $|\mathrm{Stab}_G(x_0)|$ elements. Therefore, $|\Omega| = \sum_{x \in X} |\mathrm{Stab}_G(x_0)|$. But this is $|\mathrm{Stab}_G(x_0)||G/\mathrm{Stab}_G(x_0)| = |G|$. Thus $|G| = \sum_{g \in G} f(g)$ and so in this case $\int_G f = 1$. Now suppose $X = \cup_{i \in I} X_i$, a disjoint union of orbits. Then $\int_G f = \sum_{i \in I} 1 = |I|$. $\quad\square$

We need one more lemma.

Lemma 7.2.3. *Assume G acts faithfully on X. Then $\int_G f^2 \geq 2$.*

Proof. Consider the action of G on $X \times X$ given by $g \cdot (x, y) = (g(x), g(y))$, where $g \in G$ and $(x, y) \in X \times X$. This is clearly an action. Now for $g \in G$, $f(g)^2$ is the number of points $(x, y) \in X \times X$ fixed by g. This is because the cardinality of a product of sets $S \times T$ is $|S||T|$.

Let Δ be the diagonal. Then both Δ and its complement are G-invariant. The first statement follows from the definition of the action, while the second follows from the fact that G acts faithfully on X. Therefore, there are at least two orbits here. Hence by Lemma 7.2.2 $\int_G f^2 \geq 2$. $\quad\square$

Proof of Jordan's Theorem.

Proof. Let $|G| = n$ and $G_0 = \{g \in G : f(g) = 0\}$. That is, G_0 is the set of all $g \in G$ which fix no point of X. Then, for $g \in G \setminus G_0$ we have $1 \leq f(g) \leq n$. Hence $f(g) - 1 \geq 0$ and $f(g) - n \leq 0$, so that $\psi(g) = (f(g) - 1)(f(g) - n) \leq 0$. Hence,

$$\int_{G \setminus G_0} \psi(g) \leq 0.$$

Since $\int_G = \int_{G \setminus G_0} + \int_{G_0}$, $\int_G \psi \leq \int_{G_0} \psi$. Now the right hand side of this inequality is $n|G_0|$. Thus,

$$\int_G (f(g)^2 - (n+1) \int_G f(g) + n \leq n|G_0|.$$

Since G acts transitively, Lemma 7.2.3 tells us $\int_G f(g)^2 \geq 2$, while Lemma 7.2.2 tells us $\int_G f(g) = 1$. Hence $2 - (n+1) + n \leq n|G_0|$ and so $|G_0| \geq \frac{1}{n} > 0$. $\quad\square$

Corollary 7.2.4. *If H is a proper subgroup of a finite group G, there is a conjugacy class C_g of G disjoint from H.*

Proof. Consider the action of G on $G/H = X$ by left translation. Then this action is transitive and if $g(g_1 H) = g(g_2 H)$, then $(gg_2)^{-1}gg_1 = g_2^{-1}g_1 \in H$. Thus $g_1 H = g_2 H$ and G acts faithfully. By Theorem 7.2.1 there is a $g \in G$ which fixes no point of X. That is, C_g is disjoint from H. □

Our final corollary here, asserted in [91], which we shall not prove, shows an important distinction between finite groups and compact connected Lie groups. Compare, Theorem 3.4.5, where the conjugates of maximal torus fill out G. For other similar distinctions between finite groups and compact connected Lie groups see [77].

Corollary 7.2.5. *If H is a proper subgroup of a finite group G with a disjoint conjugacy class C of G, then there exist two distinct characters of G which agree on restriction to H.*

7.3 A Fixed Point Theorem in Complex Analysis

As a final result in this chapter we mention the following theorem which gives a unique fixed point for a holomorphic function on the open unit disk, D. Although getting a fixed point could be accomplished by Brouwer's theorem, here because we have a *holomorphic* function, the fixed point actually lies in the interior, D, and is unique. This is Corollary 4.3.5 of Moskowitz [78]. Its proof uses the maximum modulus theorem and Rouché's theorem.

Theorem 7.3.1. *Let f be a holomorphic function defined on a domain containing the closed unit disk, $\bar{D}$. If $|f(z)| < 1$ on $\bar{D} \setminus D$, then f has a unique fixed point in D.*

Chapter 8

A Fixed Point Theorem in Set Theory

8.0.1. Tarski proved in 1955 that every complete lattice has the fixed point property. Later, the same year, Anne Davis proved the converse, i.e. that every lattice with the fixed point property is complete (see [29]). In order to prove this result we will need the following basic definition.

Definition 8.0.2. A *partially ordered set* is a set P equipped with a relation $\preceq$ satisfying:

- $a \preceq a$ for any $a \in P$ (*reflexive*).

- If $a \preceq b$ and $b \preceq a$, then $a = b$ (*symmetric*).

- If $a \preceq b$ and $b \preceq c$, then $a \preceq c$ (*transitive*).

Before stating our fixed point theorem we shall need to define some terms.

Definition 8.0.3. If $(P, \preceq)$ is a partially ordered set and $Q \subseteq P$, an element $p \in P$ is called an *upper bound* of Q if $q \preceq p$ for all $q \in Q$. An upper bound $p \in P$ is called a *least upper bound* of Q, (l.u.b.), if $p \preceq s$ for any other upper bound s of Q. The notions of *lower bound* and *greatest lower bound* (g.l.b.) are defined dually.

Evidently, if they exist, least upper bounds (and greatest lower bounds) are unique.

Definition 8.0.4. A *lattice* is a partially ordered set $(P, \preceq)$ in which every pair of elements x, $y \in P$ has a least upper bound, denoted by $x \vee y$, and a greatest lower bound, denoted by $x \wedge y$.

For example, and we leave this for the reader to check, given any set X its power set $\mathcal{P}(X)$ (the set of all its subsets) is a partially ordered set under inclusion $\subseteq$. Moreover, it is a lattice with least upper bound, $\cup$, and greatest lower bound, $\cap$.

We now define the environment in which the Tarski fixed point theorem works. Namely, that of a complete lattice.

Definition 8.0.5. A *complete lattice* is a partially ordered set $(P, \preceq)$ with the property that every subset $Q \subseteq P$ (finite or infinite) has a least upper bound denoted by $\bigvee Q$, and a greatest lower bound denoted by $\bigwedge Q$. $\bigvee P$ is called the *top* element and is usually denoted by $\top$ while $\bigwedge P$ is called the *bottom* element and is denoted by $\perp$.

For example, if X is any set, then $\mathcal{P}(X)$ is a complete lattice under $\subseteq$. Indeed if $A := \{A_i, \ i \in I\}$ is a subset of $\mathcal{P}(X)$, then its least upper bound is

$$\bigvee_i A = \bigcup_{i \in I} A_i$$

and its greatest lower bound

$$\bigwedge_i A = \bigcap_{i \in I} A_i.$$

Of course a complete lattice is a lattice, since every pair of elements has a least upper and greatest lower bound. However the converse is not true in general. We ask the reader to check this on the set $\mathbb{Z}^+$ of all natural numbers with the usual ordering.

Definition 8.0.6. Given a partially ordered set P, and a, $b \in P$, with $a \preceq b$, we call an *interval* in P, the set $[a, b] \subseteq P$ defined by

$$[a, b] := \{x \in P \mid a \preceq x \preceq b\}.$$

Intervals are important because of the following proposition:

Proposition 8.0.7. *If L is a complete lattice, then for any a, $b \in L$, with $a \preceq b$, $[a, b]$ is a complete lattice.*

Proof. Consider any subset $S \subseteq [a, b]$. Since L is complete, S has a least upper bound $s_1 \in L$. We will show that $s_1 \in [a, b]$. Indeed, since b is an upper bound for $[a, b]$, b is also an upper bound for S. Hence $s_1 \preceq b$. In addition, since a is a lower bound for $[a, b]$, a is a lower bound for S. Thus for any $x \in S$, we get $a \preceq x \preceq s_1$. Therefore $a \preceq s_1 \preceq b$, so $s_1 \in [a, b]$. Similarly, the greatest lower bound s_0 of S is in $[a, b]$. □

Definition 8.0.8. Given a partially ordered set, and in particular a complete lattice L, we say a function $f : L \to L$ is *order preserving* if $x \preceq y$ implies $f(x) \preceq f(y)$.

Theorem 8.0.9. (Knaster-Tarski Theorem.) *If $(L, \preceq)$ is a complete lattice and $f : L \to L$ is an order preserving function, then f has a fixed point. In fact, the set of fixed points of f is again a complete lattice*[1].

Proof. Let L be a complete lattice and $f : L \to L$ be an order preserving map and F the set of fixed points of f. Consider the set

$$B = \{x \in L \ : \ f(x) \succeq x\}.$$

Evidently, $B \neq \emptyset$, since $\bigwedge L \in B$ (L as a complete lattice has a minimum). Now, consider the element $b = \bigvee B$. Since $b \succeq x$ for all $x \in B$, and f is order preserving,

$$f(b) \succeq \bigvee_{x \in B} f(x) \succeq \bigvee_{x \in B} x = b.$$

Therefore, $b \in B$. But this implies that $b \preceq f(b)$, and since f is monotone $f(b) \preceq f^2(b)$. Hence $f(b) \in B$, so $f(b) \preceq b$. We conclude that $f(b) = b$, i.e. b is a fixed point. It is clear that $F \subseteq B$, and since b is an

---

[1]Knaster and Tarski proved a particular case of this theorem in 1927. Namely, for any set X, and $\subseteq$-increasing map $f : P(X) \to P(X)$, f has a fixed point.

upper bound of B, it must be an upper bound of F. In other words, b is the largest fixed point.

A similar argument shows that the greatest lower bound of

$$A = \{x \in L \ : \ x \succeq f(x)\}$$

is also a fixed point. Call this fixed point a (i.e. f has at least two fixed points). Again, $F \subseteq A$, and since a is a lower bound of A, it must be a lower bound of F, i.e. a is the smallest fixed point.

It remains to prove that F is, in fact, a complete lattice. In other words, we have to show if $S \neq \emptyset$ is any subset of F, then S has a least upper bound and a greatest lower bound, both in F. By assumption L is a complete lattice, hence S has a least upper bound $s_1 \in L$. But $S \subseteq L$, so $s_1 \preceq b$. Now, consider the interval $[s_1, b]$. By Proposition 8.0.7, $[s_1, b]$ is a complete lattice. We will prove that f maps $[s_1, b]$ into itself. This will imply (by the same argument used at the beginning of the proof) that f has a smallest fixed point in $[s_1, b]$, which we denote by $\bar{s}$. We claim that $\bar{s}$ is the least upper bound of S in F. Indeed, $\bar{s}$ is an upper bound since it is in $[s_1, b]$, and it is the least upper bound of S in F, because it is $\preceq$ to any other fixed point in $[s_1, b]$, and therefore $\preceq$ to any fixed point that is an upper bound of S. If s_1 is a fixed point then $\bar{s} = s_1$.

Now, to see that f maps $[s_1, b]$ into itself, take any $x \in [s_1, b]$. Since b is a fixed point of f, and by assumption f is order preserving, the fact that $x \preceq b$, implies $f(x) \preceq f(b) = b$. Similarly, $x \in [s_1, b]$ implies $x \succeq s_1$. Hence $x \succeq s$ for any $s \in S$. Thus, since f is an order preserving map and every $s \in S$ is a fixed point, $f(x) \succeq f(s) = s$ for any $s \in S$. Therefore $f(x)$ is an upper bound of S, and since s_1 is the least upper bound, $f(x) \succeq s_1$. Hence, $s_1 \preceq f(x) \preceq b$, i.e. $f(x) \in [s_1, b]$.

The proof that S has a greatest lower bound in F uses a similar argument, which we ask the reader to supply. Therefore, S has a least upper bound and a greatest lower bound in F. Thus, F is a complete lattice. □

Remark 8.0.10. The converse of Tarski's theorem is also true. In other words, for a lattice L, if any order preserving function has a fixed point, then L is complete.

We conclude with some applications of Tarski's theorem.

Corollary 8.0.11. *Let x, y be in $\mathbb{R}$ with $x \leq y$. Since the closed interval $[x, y]$ is a complete lattice relative to $\leq$, every monotone increasing map $f : [x, y] \to [x, y]$ must have a greatest fixed point and a least fixed point. Note here f need not be continuous.*

Finally, we come to the fundamental Schröder-Cantor-Bernstein theorem.

Theorem 8.0.12. (Schröder-Cantor-Bernstein).[2]   *Let A and B be sets and $f : A \to B$ and $g : B \to A$ be injections, then there is a bijection $h : A \to B$.*

Proof. As we know, $(\mathcal{P}(A), \subseteq)$ ordered by set theoretic inclusion is a complete lattice. Consider the function

$$w : \mathcal{P}(A) \to \mathcal{P}(A), \quad S \mapsto w(S) := g(f(S)^c)^c.$$

The function w is monotone increasing. To see this, let C, D be in $\mathcal{P}(A)$ with $C \subseteq D$. Then, $f(C) \subseteq f(D)$, so $f(D)^c \subseteq f(C)^c$. Hence $g(f(D)^c) \subseteq g(f(C)^c)$ and so $g(f(C)^c)^c \subseteq g(f(D)^c)^c$, which means $w(C) \subseteq w(D)$.

---

[2]The torturous history of this theorem is the following:

- 1887 Richard Dedekind proves the theorem, but does not publish it.

- 1895 Georg Cantor states the theorem in his first paper on set theory and transfinite numbers (as an easy consequence of the linear order of cardinal numbers which he was going to prove later). Here Cantor relied on the axiom of choice by inferring the result as a consequence of the well-ordering theorem. Note that the argument given below shows the result is independent of the axiom of choice.

- 1896 Ernst Schröder announces a proof (as a corollary of a more general statement).

- 1897 Felix Bernstein, a young student in Cantor's Seminar, presents his proof.

- 1897 After a visit by Bernstein, Dedekind independently proves it a second time.

- 1898 Bernstein's proof is published by Emile Borel in his book on function theory. (Communicated by Cantor at the 1897 congress in Zurich.)

Since w is an order preserving map on the complete lattice $\mathcal{P}(A)$, by Tarski 8.0.9, it must have a fixed point H. Now, we define the map $h : A \to B$ such that $h|_H = f$, and $h|_{H^c} = g^{-1}$.

It remains to show h is well defined and a bijection. By definition, $H = w(H) = g(f(H)^c)^c$. Thus $H^c = g(f(H)^c)$. Therefore, any element of H^c is in the image of g. This means the map g^{-1} is well defined on it and thus h is itself well defined.

To see h is injective, we observe h is certainly injective when restricted to H or H^c. What remains is to check what happens if $h(x) = h(y)$ with $x \in H$ and $y \in H^c$. But this cannot occur, since $H = g(f(H)^c)^c$ and the injectivity of g implies $g^{-1}(H^c) \subseteq f(H^c)$. Finally, to see that h is surjective, we remark that by definition, $H = g(f(H)^c)^c$. Hence $g^{-1}(H^c) = f(H)$. $\qquad\square$

Afterword

A number of open problems have arisen in course of the exposition and we would be very pleased if this book inspires people to work on some of them. These are to be found principally in Chapters 4 and 5.

In Chapter 4, Weinstein's theorem (see 4.3.4) has been conjectured to hold for arbitrary diffeomorphisms and not just for isometries, or conformal maps.

In Chapter 5, Arnold's conjecture of the 1960's has been partially resolved. This is the celebrated Conley-Zehnder theorem (see [28]). We also have the question of Smale seeking classification of all compact manifolds admitting Anosov diffeomorphisms. A possible answer to this could be the settling of the conjecture of Gorbatsevich: If a compact manifold M admits an Anosov diffeomorphism, then M is homeomorphic to a nilmanifold, or an infra-nilmanifold.

A final question is: For which compact, connected manifold M whose fundamental group has no elements of finite order, is it true that $N(f) = |L(f)|$ for any $f \in \mathrm{Diff}(M)$? In particular, for which M of the form G/Γ is this true where G is a connected Lie group and Γ a cocompact discrete subgroup?

During the course of working on this book the project has expanded and as a result the authors have become knowledgable in some areas of mathematics with which they were not familiar at the outset, thus providing both the readers *and writers* with an opportunity to expand their horizons.

Bibliography

[1] H. Abbaspour, M. Moskowitz, *Basic Lie Theory*, World Scientific Publishing Co., 2007.

[2] J.F. Adams, *Lectures on Lie Groups*, W.A. Benjamin, Inc., New York, 1969.

[3] P. Alexandroff, H. Hopf, *Topologie*, J. Springer, Berlin, 1935, reprinted by Chelsea, New York, 1965.

[4] D. Anosov, *The Nielsen numbers of maps of nil-manifolds*, AMS Translation Series: Russian Math. Surveys, vol. **40**, no. 4, 1985, (149-150).

[5] E. Artin, B. Mazur, *On periodic points*, Ann. of Math. vol. **81**, no. 2, 1965, (82-99).

[6] M.F. Atiyah, R. Bott, *A Lefschetz fixed point formula for elliptic complexes, I*, Ann. of Math., vol. **86**, 1967, (374-407).

[7] M.F. Atiyah, R. Bott, *A Lefschetz fixed point formula for elliptic complexes, II*, Ann. of Math., vol. **88**, 1968, (451-491).

[8] H. Azad and I. Biswas *On the Conjugacy of Maximal Unipotent Subgroups of Real Semisimple Lie Groups*, J. of Lie Theory, vol. **22**, 2012, (883-886).

[9] M. Baake, E. Lau, V. Paskunas, *A note on the dynamical zeta function of general toral endomorphisms*, Monatshefte für Mathematik, vol. **161**, no. 1, 2010, (33-42).

[10] M. Baake, J.A.G. Roberts, A. Weiss, *Periodic orbits of linear endomorphisms on the 2-torus and its lattices*, Nonlinearity vol. **21**, no. 10, 2008, (2427-2446).

[11] C. Bessaga, *On the converse of the Banach fixed-point principle*, Colloq. Math., vol. **7**, 1959, (41-43).

[12] G.D. Birkhoff, *Proof of Poincaré's geometric theorem*, Trans. AMS vol **14**, 1913, (14-22).

[13] C. Bohm and B. Wilking, *Manifolds with positive curvature operators are space forms*, Annals of Math. vol. **187**, 2008, (1079-1097).

[14] A. Borel and J.P. Serre, *Sur certaines sous-groupes des groupes de Lie compacts*, Comment. Math. Helv., vol. **27**, 1953, (128-139).

[15] A. Borel, *Linear Algebraic Groups*, 2nd enlarged edition, G.T.M. **126**, Springer-Verlag, New York, 1991.

[16] R. Bott, L.W. Tu, *Differential Forms in Algebraic Topology*, Springer-Verlag, New York, 1982.

[17] A. Brauer, *A new proof of theorems of Perron and Frobenius on non-negative matrices*, Duke Math. J., vol. **24**, 1957, (367-378).

[18] G.E. Bredon, *Topology and Geometry*, G.T.M. **139**, Springer Verlag, 3rd edition, 1995.

[19] S. Brin and L. Page, *The Anatomy of a Large-Scale Hypertextual Web Search Engine*, Computer Science Department, Stanford University, Stanford, CA 94305, USA, 1998.

[20] R. Brooks, R. Brown, J. Pak, D. Taylor, *Nielsen Numbers of Maps of Tori*, Proc. A.M.S., vol. **52**, 1975, (398-400).

[21] R.F. Brown, *The Lefschetz Fixed Point Theorem*, Scott, Foresman and Co., Glenview, IL, 1970.

[22] L.E.J. Brouwer *Uber Abbildung von Mannigfaltigkeiten*, Math. Annalen, 97-115, (1912).

[23] E. Cartan, *Groupes simples clos et ouverts et geometrie Rieman-nienne*, Journal de Math. Pures et Appliqués **8**, 1929, (1-33).

[24] G. H. Choe, *A Topological Proof of the Perron-Frobenius Theorem*, Comm. Korean Math. Soc., vol. **9**, 1994, no. 3, (565-570).

[25] G. Choquet, *Lectures on Analysis, vol II*, W.A. Benjamin, Inc., New York, 1969.

[26] H. Chu, S. Kobayashi, *The Automorphism Group of a Geometric Structure*, Trans. Amer. Math. Soc., vol. **113**, 1963, (141-150).

[27] Ana Cannas da Silva, *Lectures on Symplectic Geometry*, Lecture Notes in Mathematics, vol. **1764**, Springer, 2008.

[28] C. Conley, E. Zehnder, *The Birkhoff-Lewis fixed point theorem and a conjecture of V.I. Arnold*, Invent. Math., vol. **73**, 1983, (33-49).

[29] A. Davis, *A Characterization of Complete Lattices*, Pacific Journal of Math, vol.**5**, 1955, (311-319).

[30] G. Debreu, I. N. Herstein, *Nonnegative square matrices*, Econometrica, 21, 1953 , (514-518).

[31] P. Delorme, *1-cohomologie des representations unitaires des Groupes de Lie semi-simple et resolubles*, Bull. Soc. Math. France, vol. **105**, 1977, (281-336).

[32] Manfredo do Carmo, *Riemannian Geometry*, Birkhauser, 1992.

[33] N. Dunford and J.T. Schwartz *Linear Operators, Part I*, Interscience, New York, 1958.

[34] P. Eberlein, *Geometry of Nonpositively Curved Manifolds*, Chicago Lectures in Math., U. of Chicago Press, Chicago IL, 1996.

[35] J. England, *The zeta function of toral endomorphisms*, Proc. Amer. Math. Soc., vol. **34**, 1972, (321-322).

[36] J. Erven, B.-J. Falkowski, *Low Order Cohomology and Applications*, Lecture Notes in Mathematics, vol. **877**, Springer-Verlag, Berlin, 1980.

[37] I. Farmakis, *Cohomological Aspects of Complete Reducibility of Representations*, LAP LAMBERT Academic Publishing, Saarbrucken, 2010.

[38] B. Fortune, A. Weinstein, *A symplectic fixed point theorem for complex projective spaces*, Bull. Amer. Math. Soc., vol. **12**, 1985, (128-130).

[39] M. Frame, B. Johnson, J. Sauerberg, *Fixed Points and Fermat: A Dynamical Systems Approach to Number Theory*, Amer. Math. Monthly, vol. **1**, no. 5, 2000, (422-428).

[40] J. Franks, *Anosov Diffeomorphisms on Tori*, Trans. Amer. Math. Soc., vol. **145**, 1969, (117-124).

[41] J. Franks, *Geodesics on S^2 and periodic points of annulus homeomorphisms*, Invent. Math., vol. **108**, 1992, (403-418).

[42] J. Franks, *Area Preserving Homeomorphisms of Open Surfaces of Genus Zero*, New York Jour. of Math., vol. **2**, 1996, (1-19).

[43] G. Frobenius, *Über Matrizen aus positiven Elementen, I and II*, Sitzungsberschte der koniglish preussischen Akademie der Wissenschaften zu Berlin, 1908, (471-476), and 1909, (514-518).

[44] G. Frobenius, *Über Matrizen aus nicht negativen Elementen*, Sitzungsberschte der koniglish preussischen Akademie der Wissenschaften zu Berlin, 1912, (456-477).

[45] H. Furstenberg, *The Structure of Distal Flows*, Amer. J. Math., vol. **85**, 1963, (477-515).

[46] V. V. Gorbatsevich, *On Algebraic Anosov Diffeomorphisms on Nilmanifolds*, Sibirian Mathematical journal, vol. **45**, no 5, 2004, (821-839).

[47] F.P. Greenleaf, *Invariant Means on Topological Groups*, Van Nostrand Mathematical Studies, vol. **16**, 1969.

[48] V. Guillemin, A. Pollack, *Differential Topology*, Prentice Hall, Inc., 1974.

[49] J.L.G. Guirao, J. Llibre, *On the Lefschetz periodic point free continuous self-maps on connected compact manifolds*, Topology and its Applications, vol. **158**, no. 16, 2011, (2165-2169).

[50] J.L.G. Guirao, J. Llibre, $\mathcal{C}^{\infty}$ *self-maps on* $\mathbb{S}^n$, $\mathbb{S}^n \times \mathbb{S}^m$, $\mathbb{C}\mathbf{P}^n$ *and* $\mathbb{H}\mathbf{P}^n$ *with all their periodic orbits hyperbolic*, (English), Taiwanese J. Math. vol. **16**, no. 1, 2012, (323-334).

[51] A. Hatcher, *Algebraic Topology*, Cambridge University Press; 1st edition, 2001.

[52] M. Hausner, J. Schwartz, *Lie groups, Lie Algebras*, Gordon and Breach, New York, 1968.

[53] D. R. Heath-Brown *Fermat's two-squares theorem*, Invariant, 1984, (3-5).

[54] S. Helgason, *Differential Geometry, Lie Groups and Symmetric Spaces*, Academic Press, New York, 1978.

[55] S. Helgason, *Geometric Analysis on Symmetric Spaces*, Amer Math. Society, 1994.

[56] J.L. Heitsch, *The Lefschetz Principle, Fixed Point Theory, and Index Theory*, 2009.

[57] G.P. Hochschild, *The Structure of Lie Groups*, Holden Day, San Francisco, 1965.

[58] H. Hopf, *Vektorfelder in n-dimensionalen Mannigfaltigkeiten*, Math. Ann., vol. **96** 1926, (225-250).

[59] G.A. Hunt, *A Theorem of E. Cartan*, Proc. Amer. Math. Soc., **7**, 1956, (307-308).

[60] J. Jachymski, *A short proof of the converse to the contraction principle and some related results*, Topological Methods in Nonlinear Analysis, Journal of the J. Schauder Center, vol. **15**, 2000, (179-186).

[61] N. Jacobson, *A note on automorphisms and derivations of Lie algebras* Proc. AMS **6**, 1955, (281-283).

[62] Jürgen Jost, *Riemannian Geometry and Geometric Analysis*, Springer-Verlag, berlin, 1995.

[63] S. Kakutani, *Topological properties of the unit sphere in Hilbert space*, Proc. Imp. Acad. Tokyo, vol. **19**, 1943, (269-271).

[64] S. Kobayashi, K. Nomizu, *Foundations of Differential Geometry*, vol. II, John Wiley and Sons, Inc., New York, 1969.

[65] S.R. Komy, *On the first cohomology group for simply connected Lie groups*, J. Phys. A: Math. Gen. **18**, Issue 8, 1985, (1159-1165).

[66] P. Le Calvez, J. Wang, *Some Remarks on the Poincaré-Birkhoff Theorem*, Proc. AMS, vol. **138**, 2010, (703-715).

[67] A. Malcev, *On a class of homogeneous spaces*, Amer. Math. Soc. Translation Series, vol. **39**, 1951.

[68] W.S. Massey, *Algebraic Topology: An Introduction*, Springer, GTM 56, 1977.

[69] J. Milnor, *On fundamental groups of complete affinely flat manifolds*, Advances in Math., vol. **25**, 1977, (178-187).

[70] J. Milnor, *Morse Theory*, Princeton University Press, 1963.

[71] J. Milnor, *Topology from the Differentiable Viewpoint*, Princeton University Press, 1965.

[72] M. Moskowitz, *A note on automorphisms of Lie Algebras*, Atti della Accademia Nazionale dei Lincei, ser. 8, vol. **51**, 1971, (1-4).

[73] M. Moskowitz, *Some remarks on Automorphisms of Bounded Displacement and Bounded Cocycles.*, Monats Heft. Math., vol. **85**, 1978, (323-336).

[74] M. Moskowitz, *Complete reducibility and Zariski density in linear Lie groups*, Mathematische Zeitschrift, vol. **232**, 1999, (357-365).

[75] M. Moskowitz, *On the surjectivity of the exponential map in certain Lie groups*, Annali di Matematica Pura ed Applicata, **Serie IV-Tomo CLXVI**, 1994, (129-143).

[76] M. Moskowitz, *Correction and addenda to: On the surjectivity of the exponential map in certain Lie groups*, Annali di Matematica Pura ed Applicata, **Serie IV-Tomo CLXXIII**, 1997, (351-358).

[77] M. Moskowitz, *On a Certain Representation of a Compact Group*, J. of Pure and Applied Algebra, vol. **36**, 1985, (159-165).

[78] M. Moskowitz, *A Course in Complex Analysis in One Variable*, World Scientific Pub. Co. 2002.

[79] M. Moskowitz and F. Paliogianis, *Functions of Several Real Variables*, World Scientific Pub. Co. 2011.

[80] M.R. Murty, N. Thain, *Pick's Theorem via Minkowski's Theorem*, Amer. Math. Monthly, vol. **114**, No. 8, 2007, (732-736).

[81] M. Nakahara *Geometry, Topology and Physics*, Taylor and Francis, Second Edition 2003.

[82] I. Namioka, *Kakutani-type fixed point theorems: A survey*, J. of Fixed Point Theory and Applications, 2011, (1-23).

[83] A.L. Onishchik, E. B. Vinberg, *Lie groups and algebraic groups*, Springer-Verlag, Berlin etc, 1990.

[84] O. Perron, *Zur Theorie der Matrizen*, Mathematische Annalen, vol. **64**, 1907, (248-263).

[85] P. Petersen, *Riemannian Geometry*, GTM 171, Second Edition, Springer-Verlag, 2010.

[86] Henri Poincaré, *Sur un théorème de géometrie*, Rend. Circolo Mat. Palermo 33, 1912, (375-407).

[87] J. W. Robbin, *Topological Conjugacy and Structural Stability for Discrete Dynamical Systems*, Bull. Amer. Math. Soc., vol. **78**, no. 6, 1972, (923-952).

[88] H. Samelson, *On the Perron-Frobenius theorem*, Michigan Math., vol. **4**, 1956, (57-59).

[89] H. Samelson, *Note on Vector Fields in Manifolds*, Proceedings of the American Mathematical Society, vol. **36**, no. 1, November 1972, (273-275).

[90] M. Searcoid, *Elements of Abstract Analysis*, Springer-Verlag, 2002.

[91] J.P. Serre, *On a Theorem of Jordan*, Bull. AMS, vol **40**, no. 4, (429-440).

[92] S. Smale, *Differentiable dynamical systems*, Bull. Amer. Math. Soc., vol. **73**, 1967, (747-817).

[93] J.L. Synge, *On the connectivity of spaces of positive curvature*, Quart. J. Math. (Oxford Series), vol. **7**, 1936, (316-320).

[94] S. Tabachnikov, *Geometry and Billiards*, Amer. Math. Soc., 2005.

[95] T. Tao, *Poincaré's legacies: pages from year two of a mathematical blog* , vol. **I**, **II**, AMS, 2009.

[96] A. Tarski, *A lattice-theoretical fixpoint theorem and its applications*, Pacific Journal of Mathematics, vol. **5**, 1955, (285-309).

[97] P. Tauvel and R. Yu, *Lie Algebras and Algebraic Groups*, Springer-Verlag, 2005.

[98] R. Thom, *Espaces Fibres en Spheres et Carres de Steenrod*, Ann. Sci. Ecole Norm. Sup., vol. **69**, no. 3, 1952, (109-182).

[99] J. Tits, *Free subgroups in linear groups*, J. of Algebra, vol. **20**, no. 2, 1972, (250-270).

[100] D.E. Varberg, *Pick's Theorem Revisited*, Amer. Math. Monthly, vol. **92**, 1985, (584-587).

[101] A. Weinstein, *A fixed point theorem for positively curved manifolds*, J. Math. Mech., vol. **18**, 1968, (149-153).

[102] H. Weyl, *The Classical Groups. Their Invariants and Representations*, Princeton University Press, 1939.

[103] H. Wieland, *Unzerlegbare nicht negative Matrizen*, Mathematische Zeitschrift, vol. **52**, 1950, (642-648).

[104] D. Zagier, *A One-Sentence Proof That Every Prime $p \equiv 1 \bmod 4$ is a Sum of Two Primes*, Amer. Math. Monthly, vol. **97**, no. 2, 1990, (144).

[105] E. Zehnder, *The Arnold Conjecture for Fixed Points of Symplectic Mappings and Periodic Solutions of Hamiltonian Systems*, Proc. Inter. Congress of Mathematicians, Berkeley, California, USA, 1986.

[106] R.J. Zimmer, *Ergodic Theory and Semisimple Groups*, Birkhäuser, Boston, 1984.

Index